Contents

Section 1 Tests 1 to 12, covering revision of Book 3 plus th[e following:]

Number: 10s number system. Reading and writing tenths, hundredths. Place value; decomposition of numbers. + −, x, ÷ facts; four rules. multiples, factors, square and cubic numbers; averages.

Fractions: Equivalents. Reduction to lowest terms; fractional parts of given quantities. Addition and subtraction.

Money: Notes and coins, counting and manipulation; conversion. Shopping, costings, change. Four rules; averages; decimal parts of £1.

Measures: Length, mass, capacity. Units: mm, cm, m, km; g, kg, $\frac{1}{2}$ kg; ml, l, $\frac{1}{2}$ l. Conversions; fractional parts; four rules; averages.

Approximations: To nearest whole number, 10, 100, 1000; to nearest £1 or ten; to nearest cm, m, km; kg, $\frac{1}{2}$ kg; l, $\frac{1}{2}$ l.

Time: h, min, s; fractional parts; conversions. 12-hour, 24-hour clock times, conversions. Calendar: days, months, years. Calculation of periods of time: h and min; days, months, years and months. Time, distance, speed: km/h.

Angles and shapes: Right angles, acute and obtuse angles; adjacent angles; straight angles; vertically opposite angles. Compass directions. Squares, rectangles: perimeter, area (units of measurement). Triangles: naming by sides or angles; sum of angles = 180°. Quadrilaterals, sum of angles = 360°. Circles: radius, diameter, circumference; angles at the centre.

Progress Test 1	16
Progress Test 1 Results Chart	17

Section 2 Tests 1 to 12, covering topics from Section 1 plus the following: 18

Number: 10s number system. Reading and writing six-digit whole numbers; decimal fractions: tenths, hundredths, thousandths. Place value; decomposition of numbers. Multiplication and division by 10, 100, 1000.

Fractions: Evaluations; making fractions; finding a whole given a part. Simple conversions to decimal fractions and vice versa.

Percentages: % 'out of 100'; conversions to fractions (lowest terms) and decimals. 5%, 10%, 20%, 25%, 50%, 75%, 100%. Fraction and decimal equivalents. Evaluation of percentages of given quantities.

Money: Approximations to the nearest penny. Division to the nearest penny.

Measures: Conversions in decimal form (thousandths). Mass and capacity relationships. Temperature. Celsius readings + and −.

Angles: Calculations. Interior and exterior angles (triangles); interior angles (quadrilaterals). Angles at centre of regular polygons (pentagon, hexagon, octagon).

Shapes: Squares, rectangles: perimeter, area (cm^2, m^2). A = lb, l = A/b, b = A/l. Triangles: perimeter, area (cm^2, m^2). A = bh/2, b= 2A/h, h = 2A/b. Circles: radius, diameter, circumference (π = 3.14). Cubes, cuboids: surface area; fitting cm cubes.

Progress Test 2	30
Progress Test 2 Results Chart	31

Section 3 Tests 1 to 12, covering topics from Sections 1 and 2 plus the following: 32

Number: 10s number system. Million, $\frac{1}{2}$, $\frac{1}{4}$, $\frac{3}{4}$ million, $\frac{1}{10}$ or 0.1 million.

Percentages: 10%, 60%, 80%, 30%, 70%, 90%. Evaluation of given quantities. 1% of £1, 1 m, 1 kg, 1 l. Miscellaneous percentages.

Angles and shapes: Reflex angles; angle in semicircle (90°). Area of irregular shapes (rectangles, triangles). Circles C = π d; angle at centre, length of arc. Cubes, cuboids. Volume (cm^3): V = lbh. Symmetry: 1, 2, or 3 axes of symmetry.

Check-up Tests	Number	44
	Money and Measures	45
	Fractions and Percentages	46
	Approximations, Angles and Shapes	47

Section 1   Test 1

## A		ANSWER
1	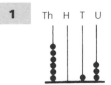 Write in words the number shown on the abacus picture.	six thousand and thirteen
2	17 + 8 + 16	41
3	965 − 60	905
4	(8 × 7) + 5	61
5	(56 − 8) ÷ 8	6
6	7/10 of 100 g	70 g
7	1¾ h = ▢ min	105 min
8	£1·45 = ▢ FIVES	29 FIVES
9	4 km 350 m = ▢ m	4350 m
10	850 g + ▢ g = 1½ kg	6500 g
11	£7·09 = ▢ pennies	709 p
12	▢ TENS + 6 TWOS = £1·82	17 TENS

## B		ANSWER
1	Write in figures the number twelve thousand and eight.	12,008
2	How many groups of 9 are there in 6 sixes?	4
3	What is the difference in pence between £⅕ and £¼?	5 p
4	How many tens are equal to 1070?	107
5	Find the total of 53p and £1·37.	£1.90
6	By how many g is ½ kg heavier than 280 g?	780 g
7	Find the cost of 9 articles at 13p each.	£1.17
8	How many mm are there in 10.7 cm?	107 mm
9	How much change from a FIFTY after spending 17p and 16p?	17 p
10	Change to 24-hour clock times (a) 9.35 a.m. (b) 8.50 p.m.	(a) 9.35 (b) 20.50
11	Find the smallest number which will divide by both 6 and 8 without a remainder.	5
12	What sum of money when multiplied by 7 equals £1·12?	16 p

## C		ANSWER
1	In a box were 48 cards. How many cards were there in 7 boxes?	336
2	In the number 7479 how many times is the 7 marked x greater than the 7 marked y?	100
3	A motorist travelled 348 km in 6 hours. Find his average speed in km/h.	58 km/h
4	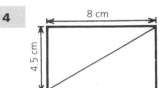 This rectangular card is cut into two equal parts along a diagonal. Find the area of each part.	18 cm
5	What are the next two numbers in this series? 1/10, 1, 10, ▢, ▢	100, 1000
6	Find the difference in cost between 12 articles at 3p each and 12 articles at 5p each.	24 p
7	A bus leaves its station at 8.40 a.m. and arrives at its destination at noon. How long does the journey take?	3 h 20 min
8	PRICE OF CLOTH (A) £3·96 per m (B) £4·18 per m — By how much is cloth (B) more expensive per m than cloth (A)?	22 p
9	The mass of fruit in a tin is 425 g. Find in kg and g the mass of the fruit in 10 tins.	4 kg 250 g
10	Through how many degrees does the minute hand of a clock turn in 10 minutes?	60°
11	A plank of wood is cut into 12 equal pieces. What fraction of the plank is 7 pieces?	7/12
12	 The diagram shows how Suki used her prize money. (a) What fraction did she spend? (b) The prize was £40. How much did she save?	(a) 3/8 (b) £25

Section 1   Test 2

A

		ANSWER
1	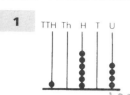 Write in words the number shown on the abacus picture.	ten thousand six hundred and four
2	27 + 6 = 20 + ☐	13
3	78 × 6	468
4	872 ÷ 8	109
5	£1·68 − 96p = ☐ p	72 p
6	27 quarters =	108
7	¼ of £4·32	£1·08
8	2080 m = ☐ km ☐ m	km 2 m 80
9	4 kg 700 g = ☐ g	4700 g
10	£10 − £0·82	£9·18
11	☐ FIFTIES + 7 FIVES = £1·85	3 FIFTIES
12	The time on the clock is 7 min slow. Write the correct time using a.m. or p.m.	5.46 pm

B

		ANSWER
1	Find the missing number. 5000 + ☐ + 7 = 5087	80
2	How many TWENTIES have the same value as 3 TENS and 1 FIFTY?	4 TWENTIES
3	Share £1·56 equally among 6 people. How much each?	26 p
4	Write 3.45 m as cm.	345 cm
5	Which of these numbers divide by 8 without a remainder? 12, 28, 32, 44, 56, 68	32, 56
6	Find the value of (a) ⅕ of £35 (b) ⅘ of £35.	(a) £7 (b) £28
7	Write 996 to the nearest 10.	1000
8	How many g in 2¼ kg?	2250 g
9	Find the change from £1 after spending 19p and 17p.	64 p
10	What is the average of 6 cm, 9 cm, 7 cm, 10 cm?	8 cm
11	Change to 12-hour clock times (a) 09.05 (b) 16.48.	(a) 9.05 am (b) 4.48 pm
12	10 articles cost £1·30. Find the cost of one article.	13 p

C

		ANSWER
1	A cricketer scored 12 runs short of a century. How many runs did he score?	88
2	3.7 × 9 = 33.3 Write the value of 3.7 × 90.	333
3	The price of fish is £6·40 per kg. What is the cost of 1 kg 500 g?	£9.60
4	How many 10-cm strips can be cut from a length of 5 m 60 cm?	56
5	There are 20 litres in the petrol tank of a car. If it is ⅓ full, how many more litres will it hold?	60 ℓ
6	Find the total value of these coins. 	£1·61
7	A school's Easter holiday started on 26th March and ended on 9th April. How many days holiday were there?	14
8	A ship is sailing due East. It changes course to SE. Through how many degrees does it turn?	45°
9	 The diagram shows how three children shared a prize of £40. How much did each child receive?	Josh £5 Sophie £15 Sunil £20
10	A train leaves at 17.50 and arrives at midnight. How long does the journey take in h and min?	6 h 10 min
11	The price of butter is increased from £1·29 to £1·34 per 500 g. How much extra is paid for 6 kg?	60 p
12	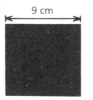 Find (a) the perimeter (b) the area of this square. (Write the unit of measurement in each case.)	(a) 36 cm (b) 15 cm²

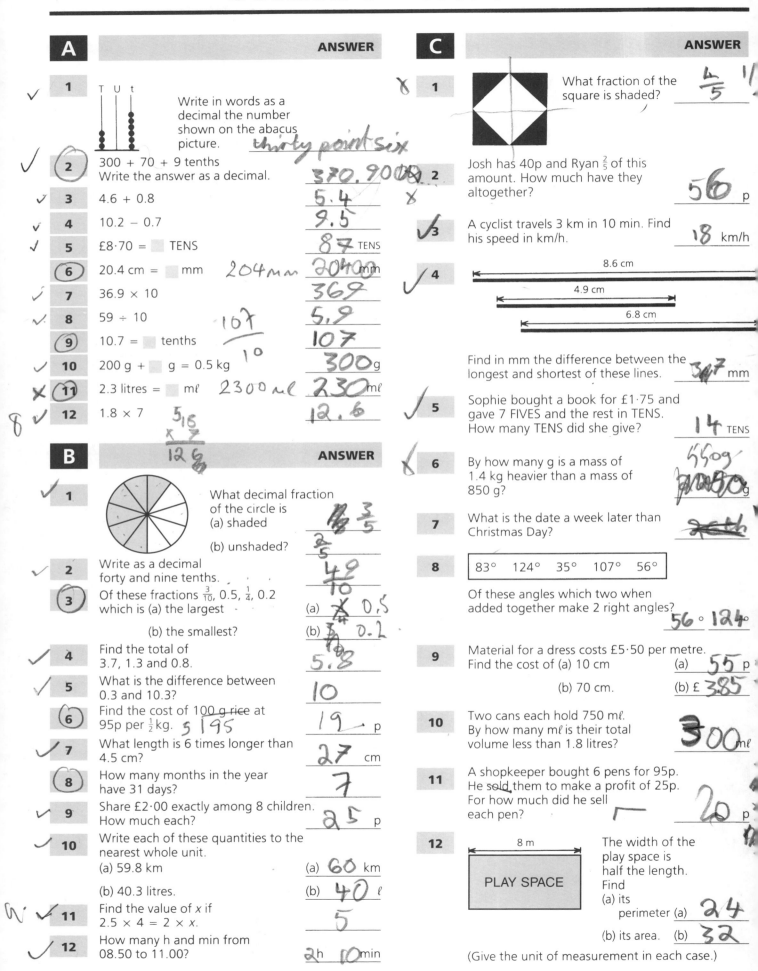

Section 1 Test 4

A

		ANSWER
1	[abacus T U t h] Write in words as a decimal the number shown on the abacus picture.	fourteen point one seven
2	8.03 = 8 + __	$\frac{3}{10}$
3	2.26 + 1.04	3.30
4	£3·04 = ▢ pence	304 p
5	248 cm = ▢ m	2.48 m
6	Write $\frac{1}{4}$ as a decimal.	0.24
7	10 − 0.88	9.12
8	5.07 × 100	507
9	Find in g (a) 0.1 of 1 kg (b) 0.7 of 1 kg.	(a) 100 g (b) 700 g
10	Write as a decimal 503 hundredths.	8.03
11	0.46 × 8	3.68
12		

B

		ANSWER
1	Write as a decimal 30 + $\frac{7}{10}$ + $\frac{4}{100}$.	30.74
2	What is the value in pence of the digit underlined? (a) £100·40 (b) £15·06	(a) 40 p (b) 6 p
3	Write in m and cm 109.46 m.	109.46 m 109.46 cm
4	Write the missing signs +, −, ×, ÷ in place of ● and ▲. 9 ● 6 = 3 ▲ 5	● ÷ ▲ ×
5	Write as a decimal 15 tenths 7 hundredths.	1.57
6	Decrease 0.5 litre by 150 mℓ.	350 mℓ
7	How many h and min from 2.25 p.m. to 4.05 p.m.?	1 h 40 min
8	Write these fractions in their lowest terms. (a) $\frac{8}{12}$ (b) $\frac{21}{24}$	(a) $\frac{2}{3}$ (b) $\frac{21}{24}$
9	Divide £15 by 6 exactly.	£
10	Find the area of a rectangle measuring 8.5 cm by 6 cm.	51 cm²
11	Divide ▢ kg by 100. Give the answer in g.	g
12	Which of these fractions are of equal value?	

C

		ANSWER
1	[10×10 grid] Write as a decimal fraction the part of the square which is (a) shaded (b) unshaded.	(a) (b)
2	Badges cost 6p each. How many can be bought for £1·56?	
3	A medicine spoon holds 5 mℓ. How many spoonfuls in $\frac{1}{4}$ litre?	
4	Find the missing number in this example. 1 7 rem. 7 9) ▢	
5	Mince costs £1·80 per $\frac{1}{2}$ kg. Find the cost of mince weighing 600 g.	£
6	A triangle has two equal sides each measuring 36 cm. Its perimeter is 1 m. Find the length of the third side.	cm
7	There were 845 people at a concert. Write this number (a) to the nearest 100. (b) to the nearest 10.	(a) (b)
8	There are two 2-place decimal numbers which are greater than 3.97 but less than 4.00. What are the numbers?	
9	Fatima collected 150 pennies for the school fund. Olivia collected three times as many. How many £1 coins did they receive in exchange?	
10	Josh's stride measures 40 cm. How many strides will he take in walking 10 m?	
11	A path measures 9 m long and 50 cm wide. Find its area in m².	m²
12	[ruler scale from A to B]	
	The line AB is drawn to a scale 1 cm to 20 cm. Find the length the line represents (a) in cm (b) in mm.	(a) cm (b) mm

Section 1 Test 5

A

		ANSWER
1	Write as a decimal $10 + \frac{3}{10} + \frac{7}{100}$	10.37
2	$(8 \times 9) + (0 \times 7)$	72
3	10.05 m = ☐ cm	1005 cm
4	250 g + ☐ g = 600 g	350 g
5	0.5 litres = 330 mℓ + ☐ mℓ	170 mℓ
6	$\frac{3}{8}$ of £40	£30
7	1.3 + 0.75	2.05
8	£2.08 − 70p = £ ☐	£2.78
9	Write $\frac{3}{5}$ as a decimal fraction.	0.6
10	3 h 40 min = ☐ min	480 min
11	40.08 × 100	4008
12	8.32 ÷ 4	2.08

B

		ANSWER
1	Write as a decimal fraction the part of the strip which is (a) shaded (b) unshaded.	(a) $\frac{3}{10}$ (b) $\frac{7}{10}$
2	What number is 7 more than 6 × 8?	55
3	(a) Name the eleventh month of the year. (b) How many days are there in that month?	(a) November (b) 30
4	Find the cost of 100 g at £1·40 per kg.	14 p
5	Find the difference between the largest and smallest of these fractions. $\frac{3}{10}, \frac{1}{2}, \frac{4}{5}, \frac{7}{10}$	0.5
6	Write 19 litres 720 mℓ to the nearest litre.	20 ℓ
7	From the sum of 3.6 and 6.2 take 5.	4.8
8	Find the area in m² of a floor measuring 8 m by 3 m 50 cm.	28 m²
9	Complete this number series. 0.3, 3, 30, ☐, ☐	300, 3000
10	Divide 30.4 cm into 8 equal parts. Find the length of each part.	3.8 cm
11	How many TWENTIES must be added to 3 TENS to equal £2·10?	4 TWENTIES
12	Of these triangles which is (a) isosceles (b) equilateral?	(a) B (b) C

C

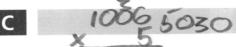

		ANSWER
1	One thousand and six people each bought 5 tickets. How many tickets was that altogether?	5030
2	How much change from 3 TWENTIES after paying for 6 eggs at £1·08 per dozen?	☐ p
3	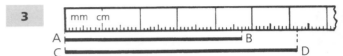 Write the length of each line in cm. AB CD	4.7 cm 6.2 cm
4	In 7 weeks Katie saves £3·15. Find her average weekly savings.	☐ p
5	A boy spent $\frac{1}{4}$ of his money on sweets and $\frac{3}{8}$ on bus fares. What fraction of his money is left?	
6	How many degrees are there in each of the equal angles at the centre of the circle?	360°
7	 How many g less than 1 kg is the total mass?	0.350 g
8	Tom's date of birth is 7th March 1985. Daniel was born exactly 4 years later. Write in figures Daniel's date of birth.	11/3/1
9	The approximate distance between two villages is given as 11 km. The actual distance is 10.7 km. Find the difference in m.	10700 m
10	Which two of these fractions are equivalent to $\frac{3}{4}$? $\frac{6}{10}, \frac{9}{12}, \frac{4}{5}, \frac{15}{20}$	$\frac{15}{20}$ $\frac{9}{12}$
11	Emily and Katie have 60p between them. Emily has 8p more than Katie. How much has each?	Emily 34p Katie 26p
12	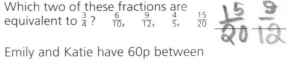 Find the missing measurement marked b.	30

8

Section 1 Test 6

A

1. 45p + 35p + £1·20   £ _____
2. 63 ÷ 8   _____ rem.
3. Write as a decimal 708 hundredths.   _____
4. $\frac{1}{2}$ kg − ___ g = 125 g   _____ g
5. £1·05 × 6   £ _____
6. 0.8 + 3.25   _____
7. How many minutes from 9.27 a.m. to 11.15 a.m.?   _____ min
8. ___ mℓ + 4050 mℓ = 5 litres   _____ mℓ
9. 60.4 ÷ 10   _____
10. $\frac{3}{5} = \frac{}{100}$ _____ /100
11. 0.5 of £17·20   £ _____
12. ∠A + ∠B + ∠C = ___°   _____ °

B

1. Write the part which is shaded
   (a) as a vulgar fraction _____
   (b) as a decimal fraction. _____
2. From 9 times 7 take 5.   _____
3. Write the 24-hour clock time for 12 minutes before midnight.   _____
4. Find the cost of 20 cm at 75p per m.   _____ p
5. Write 9 kg 870 g to the nearest $\frac{1}{2}$ kg.   _____ kg
6. Find the difference between 3.8 litres and 6 litres.   _____ ℓ
7. What length in cm is $\frac{1}{5}$ of 3 m?   _____ cm
8. Find the total of $2\frac{1}{4}$, $3\frac{5}{8}$ and 5.   _____
9. How many TWOS are worth £2·48?   _____ TWOS
10. A square has sides measuring 10 cm. Find (a) its perimeter (a) _____
    (b) its area. (b) _____
11. Complete the number series. 0.01, 0.1, 1, ___, ___   _____
12. Which of these triangles is
    (a) a right-angled triangle _____
    (b) an acute-angled triangle _____
    (c) an obtuse-angled triangle? _____

C

1. Find the total of $\frac{3}{4}$ kg, 400 g and 200 g. Write the answer as kg and g.   _____ kg _____ g
2. What is the value of the digit underlined in each of these numbers?
   (a) 6037   (a) _____
   (b) 49.08   (b) _____
3. A boy walked at an average speed of 6 km/h. How many hours will he take to walk 15 km?   _____ h
4. Samina bought 8 sweets at 4p each. How much change had she from a FIFTY?   _____ p
5. How many degrees are there in the angle marked A?   _____ °
6. The population of a small town is 8968. Write the number
   (a) to the nearest 1000   (a) _____
   (b) to the nearest 100.   (b) _____
7. Find the smallest number which can be added to 40 to make a number which is exactly divisible by 7.   _____
8. 5 balloons cost 45p. Find the cost of 3 balloons.   _____ p
9. 6 children each had an equal share of a sum of money. They each received 18p and there was 2p left over. Find the sum of money.   £ _____
10. Find the value in m of x if $\frac{x}{10} = 1.6$ m.   _____ m
11. What liquid measure is equal to 0.1 of 20 litres?   _____ ℓ
12. Find the length of
    (a) the side AB   (a) _____ cm
    (b) the side BC.   (b) _____ cm

Section 1   Test 7

A

1. $\frac{7}{12} + \square = 1$
2. Write as a decimal $15 + \frac{5}{100}$.
3. $20.4 = \square$ tenths _____ tenths
4. $(8 \times 8) + 6$
5. $2 \text{ km} + \frac{1}{2} \text{ km} + \frac{1}{4} \text{ km} = \square \text{ m}$ _____ m
6. Find the missing number. $3000 + 100 + \square + 6 = 3196$
7. £2·00 − 46p = £ _____ £
8. $700 \text{ g} + \square \text{ g} = 1.5 \text{ kg}$ _____ g
9. How many thirds in $6\frac{2}{3}$? _____ $\overline{3}$
10. The time on the clock is 13 min fast. Write the correct time using a.m. or p.m. (evening)
11. 27p × 4   £ _____
12. 0.7 litres − $\square$ mℓ = 610 mℓ   _____ mℓ

B

1. $5^2 = 5 \times 5 = 25$. Find the value of (a) 6^2 (b) 10^2.
2. Add 8 hundredths to 3.04.
3. What is the difference between 90p and £3·25?   £ _____
4. How many h and min are there in 105 min?   _____ h _____ min
5. Find the cost of 7 lemons at 18p each.   £ _____
6. Write as fractions in their lowest terms. (a) $\frac{8}{20}$ (b) $\frac{25}{100}$ (a) _____ (b) _____
7. How many g must be added to 2300 g to make $2\frac{1}{2}$ kg?   _____ g
8. 9 times 4.03
9. How many degrees are there in the angle marked A? (45°, 15°)   _____ °
10. By how many mm is 8.3 cm longer than 56 mm?   _____ mm
11. Divide £3·68 into 8 equal parts. What is the value in pence of each part?   _____ p
12. $(8 \times 6) = (3 \times 6) + (x \times 6)$. Find the number x stands for.

C

1. Write in figures the number which is 70 less than ten thousand.
2. Count the value of these coins in the given order and find the total amount. (10, 10, 50, 50, 5, 5, 20, 20, 2)   £ _____
3. A car travels 26 km in 20 min. Find its speed in km/h.   _____ km/h
4. 1 litre of water has a mass of 1 kg. Find in g the mass of water in a bottle which holds $\frac{1}{4}$ litre.   _____ g
5. In the square give the size in degrees of
   (a) angle A   (a) _____ °
   (b) angle B. (b) _____ °
6. In a test, James scored 70 out of 100. What fraction, in its lowest terms, of the total did he score?
7. Chloe has £1·50 and Olivia has $\frac{3}{5}$ of this amount. How much have they altogether?   £ _____
8. How many days are there from 28th June to 9th July? Do not count the first day.
9. 5 books together have a mass of 1 kg 450 g. Find the average mass.   _____ g
10. The plot of land is 3 times as long as it is wide. Find
    (a) its length   (a) _____
    (b) its perimeter. (b) _____
    (20 m, PLOT OF LAND)
11. By how many m² is the area of the plot more than 1000 m²?   _____ m²
12. A boy doubles his savings every week for 4 weeks. In the first week he saved 20p. How much did he save in the fourth week?   £ _____

Section 1 Test 8

A

1. 6.07 = ▯ hundredths _____ hundredths
2. Write as a 24-hour clock time 16 min after midnight. _____
3. Write as a decimal $200 + \frac{3}{10} + \frac{9}{100}$. _____
4. 5.8×7 _____
5. Find the value of x if $54 \div 9 = 30 \div x$. _____
6. £1·00 − (4 TENS + 9 FIVES) _____ p
7. $40° + 65° + $ ▯ $° = 180°$ _____ °
8. $\frac{1000}{8}$ _____
9. £4·86 = ▯ TENS + 6p _____ TENS
10. ▯ mℓ × 10 = 2 litres _____ mℓ
11. £2·34 ÷ 6 _____ p
12. 425 g + 70 g + 87 g = ▯ kg _____ kg

B

1. Add 5 to 7^2. _____
2. Change these times to 12-hour clock times using a.m. or p.m.
   (a) 07.05 (a) _____
   (b) 23.20 (b) _____
3. Decrease £1·34 by 40p. _____ p
4. Write to the nearest whole number
   (a) $9\frac{7}{10}$ (b) $20\frac{2}{5}$. (a) _____ (b) _____
5. By how many m is 1850 m less than 3 km? _____ m
6. Find the total of 15 TWENTIES and 12 TENS. _____ £
7. What is the difference in mℓ between $\frac{3}{4}$ litre and 580 mℓ? _____ mℓ
8. How many g are equal to 0.3 kg? _____ g
9. How many degrees are there in each of the equal angles at the centre of the circle? _____ °
10. $\frac{1}{2}$ kg costs £1·60. Find the cost of 100 g. _____ p
11. Find the area of a square with sides of 40 cm. _____
12. Divide 23 exactly by 5. Write the answer as a decimal number. _____

C

1. By how many is 10^2 greater than 9^2? _____
2. How many
   (a) tenths are equal to 50.6 (a) _____ tenths
   (b) hundredths are equal to 40.75? (b) _____ hundredths
3. In a bag there are 13 TWOS and 14 pennies. How many TWENTIES are exchanged for these coins? _____ TWENTIES
4. 7 nails have a mass of 50 g. How many nails have a mass of $\frac{1}{2}$ kg? _____
5. Through how many degrees does the hour hand of a clock turn from noon to 4.00 p.m.? _____ °
6. Write these numbers so that the value of the figure 7 in each number is 7 hundredths.
   (a) 1607 (a) _____
   (b) 97 (b) _____
7. A right-angled triangle contains an angle of 55°. Find the size of the third angle. _____ °
8. When a barrel is $\frac{1}{5}$ full it holds 16 litres. How many litres will it hold when $\frac{1}{2}$ full? _____ ℓ
9.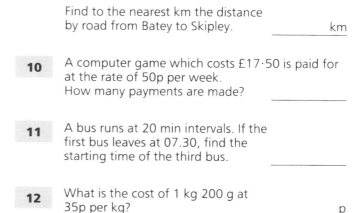
   Find to the nearest km the distance by road from Batey to Skipley. _____ km
10. A computer game which costs £17·50 is paid for at the rate of 50p per week. How many payments are made? _____
11. A bus runs at 20 min intervals. If the first bus leaves at 07.30, find the starting time of the third bus. _____
12. What is the cost of 1 kg 200 g at 35p per kg? _____ p

Section 1   Test 9

A

1. 0.8 + 0.8 + 0.8 + 0.8
2. 200 − 97
3. $\frac{3}{10}$ of 50p ___ p
4. Write as a decimal 70 + 6 + $\frac{9}{100}$.
5. 2 km 350 m = ▢ m ___ m
6. £1·71 ÷ 9 ___ p
7. 1.5 kg − 650 g = ▢ g ___ g
8. 7 FIVES − 8 TWOS = ▢ p ___ p
9. 1150 ml + ▢ ml = 2 litres ___ ml
10. 215 min = ▢ h ▢ min ___ h ___ min
11. £3·05 = ▢ TENS + 1 FIVE ___ TENS
12. 4.5 kg × 100 ___ kg

B

1. What is the value of the missing number? Thirty point three six = 30 + ▢ + $\frac{6}{100}$.
2. Write to the nearest metre 19.54 m. ___ m
3. Find the change from £5·00 after spending 89p. ___ £
4. Find the total of 0.9 and 4.38.
5. Change 8.35 p.m. to 24-hour clock time.
6. Which three coins when added together make 56p? ___ p ___ p ___ p
7. Find in mm the radius of a circle the diameter of which is 7.6 cm. ___ mm
8. Increase £1·20 by a quarter. ___ £
9. Name the pair of parallel lines in this shape.
10. How many times can 250 ml be taken from 2½ litres?
11. Find the cost of 1 kg 100 g of oranges at 50p per ½ kg. ___ £
12. Which of these decimal fractions is equal to $\frac{3}{4}$? 0.34   0.43   0.75   0.63

C

1. Arrange the digits 3, 8, 0, 5 to make
   (a) the largest possible number (a)
   (b) the smallest possible number. (b)
2. This shape has 8 equal sides.
   (a) Name the shape. (a)
   (b) How many degrees are there in ∠ A? (b) ___°
3. Share £4·76 equally among 7 children. Find how much each has. Write the answer (a) in pence (a) ___ p
   (b) in £s. (b) £ ___
4. A medicine spoon holds 5 ml. What decimal fraction of a litre is contained in 100 spoonfuls? ___ ℓ
5. Find in g (a) 0.1 kg (a) ___ g
   (b) 0.4 kg (b) ___ g
   (c) 0.8 kg. (c) ___ g
6. The perimeter of the rectangle is 96 cm. The width is 15 cm. Find its length. ___ cm
7. What is the difference between the largest and smallest of these decimal numbers? 1.1   0.98   1.12   1.06
8. The thickness of 100 sheets of card is 14.5 cm. Find the thickness in mm of (a) 10 sheets (a) ___ mm
   (b) 1 sheet. (b) ___ mm
9. Sanjay is facing NE. In which direction is he facing if he turns clockwise through 2 right angles?
10. Needles are sold at 5 for 35p. What is paid for 30 needles? £ ___
11. Tom spent $\frac{3}{5}$ of his money on sweets and $\frac{3}{10}$ on ice cream. What fraction of his money had he left?
12. Find the area of the smallest square into which the circle can be fitted.

Section 1   Test 10

A

1. Write as a decimal
   (a) 407 tenths   (a) _____
   (b) 209 hundredths.   (b) _____
2. How many days in 1 year? _____
3. $4^2 + 3^2 - 2^2$ _____
4. 1 FIFTY + 2 TWENTIES = ▇ FIVES _____ FIVES
5. 10 − 0.55 _____
6. $\frac{9}{10}$ of 1 kg = ▇ g _____ g
7. 63 ÷ 7 = 54 ÷ _____
8. $\frac{2}{5} + \frac{1}{2}$ _____
9. 0.48 m ÷ 8 = ▇ cm _____ cm
10. £2·30 − 95p = £ _____ £
11. 87 × 6 _____
12. Write as a decimal
    9 + 3 tenths + 17 hundredths. _____

B

1. Write the value of the figure underlined.
   (a) 36.0<u>9</u>   (a) _____
   (b) <u>1</u>80.6   (b) _____
2. By how many g is 1.3 kg heavier than 700 g? _____ g
3. Find the cost of 8 articles at 19p each. £ _____
4. By what length is 0.5 m less than 10.05 m? _____ m
5. Write the 24-hour clock time which is 17 min later than 13.55. _____
6. How many 12p eggs can be bought for £1·80? _____
7. Which two of these fractions when added together make a whole one?
   $\frac{5}{8}, \frac{2}{5}, \frac{1}{4}, \frac{6}{10}, \frac{1}{8}$ _____
8. Find the average of 1.6 litres, 0.8 litres, 1.2 litres. _____ ℓ
9. How may pennies remain when £2·53 is divided by 6? _____ p
10. Find in cm the diameter of a circle whose radius is 85 mm. _____ cm
11. Write each of these fractions in the lowest terms. (a) $\frac{16}{20}$ (b) $\frac{25}{30}$ (a) _____ (b) _____
12. Find the perimeter of a rectangle 12.5 cm long and 3.5 cm wide. _____ cm

C

1. What is the greatest possible remainder when a whole number is divided by 9? _____
2. How many pieces of ribbon each 4.5 cm long can be cut from a length of $4\frac{1}{2}$ m? _____
3. A bag of rice having a mass of 200 g costs 40p. Find the price per kg. £ _____
4. What is the 12-hour clock time which is $1\frac{1}{4}$ h earlier than 13.05? Use a.m. or p.m. _____
5. Eggs are packed in boxes in layers of 20. If there are 4 layers in each box and 5 boxes, find the total number of eggs. _____
6. £
   8·60
   *·**
   2·90
   17·00
   What is the missing sum of money? £ _____
7. Find the size in degrees of ∠A _____ °
   (145°, with angles A, B, C) ∠B _____ °
   ∠C. _____ °
8. 15 × 36 = 540. By how many is 16 × 36 more than 540? _____
9. A 2-litre can is $\frac{7}{10}$ full. How many more mℓ are required to fill it? _____ mℓ
10. A map is drawn to a scale 1 cm to 10 km. Find the actual distance represented by 54 mm. _____ km
11. In a collection there were 308 FIVES which were packed into £5 bags.
    (a) How many bags were filled? (a) _____
    (b) What is the value in pence of the coins left over? (b) _____ p
12. (80 cm × 30 cm rectangle) How many tiles each 10 cm square are needed to cover the surface? _____

Section 1 Test 11

A

		ANSWER
1	(7 × 8) + 5	
2	2.86 = ▨ tenths + 6 hundredths	tenths
3	4/5 = ▨/100	/100
4	£2·04 × 8	£
5	How many min from 11.40 a.m. to 1.10 p.m.?	min
6	0.6 + 0.18 + 5.4	
7	4 kg 50 g = ▨ g	g
8	70p × 2½	£
9	504 mm = ▨ cm	cm
10	∠A + ∠B + ∠C + ∠D = ▨°	°
11	0.3 litre = ▨ mℓ	mℓ
12	£9·00 ÷ 100 = ▨ p	p

B

		ANSWER
1	Take 0 times 9 from the product of 8 and 6.	
2	5 packets of different sizes have a total mass of 800 g. Find their average mass.	g
3	Decrease 72p by 1/8.	p
4	In a leap-year February has 29 days. How many days in a leap-year?	
5	Write 9 litres 700 mℓ to the nearest ½ litre.	ℓ
6	How much change from a £5 note after first spending £3 and then 56p?	£
7	Which of these fractions equal 2/3? 4/9 8/12 16/20 10/15	
8	What number when multiplied by 3 gives 207 for the answer?	
9	Find the cost of 2 m 20 cm at 25p per m.	p
10	What number must be added to 0.37 to make 4.57?	
11	17 × 7 = 119. Write the answer to 0.17 × 7.	
12	Which of these lines are perpendicular to the line XY?	

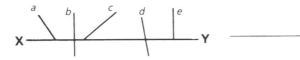

C

		ANSWER
1	Write the next two odd numbers in this series. 995, 997, 999, ▨, ▨	,
2	Bird seed costs 25p for 100 g. Find the cost of a bag of bird seed containing 1 kg 300 g.	£
3	Write each of these fractions with a denominator of 100. (a) 9/10 (b) 7/20 (c) 13/50	(a)___ (b)___ (c)___
4	The rainfall for three months of a year was 37 mm, 43 mm, 40 mm. Find the average monthly rainfall.	mm
5	A square of 20 cm side is cut along both diagonals into 4 equal triangles. What is the area of each triangle in cm²?	cm²
6	George wrote 10.6 for an answer instead of 10.06. Find his error and write it as a decimal.	
7	In this right-angled triangle what is the size in degrees of ∠A?	°
8	Find the time taken in h and min for a journey of 300 km at an average speed of 40 km/h.	h min
9	Jack needs £3·20 to buy a book. He has saved 3 FIFTIES, 4 TWENTIES and 1 TEN. How much more must he save?	p
10	A train journey from London to Leeds takes 2 h 35 min. At what time do these trains arrive at Leeds if they leave London at (a) 11.25 (b) 18.45?	(a)___ (b)___
11	1 litre of water has a mass of 1 kg. Find the mass of a bottle filled with 1.5 litres of water if the bottle has a mass of 150 g.	kg g
12	How many centimetre cubes are needed to make this block? (3 cm × 14 cm × 1 cm)	

Section 1 Test 12

A

		ANSWER
1	300 + 15 + 5000	
2	45 FIVES = £	£
3	27/100 of 1 metre = ☐ cm	cm
4	200 − 0.45	
5	The ninth month of the year is ☐.	
6	709 × 8	
7	3.7 = ☐ hundredths	hundredths
8	17p + 15p + 20p = £ ☐	£
9	140 g + ☐ g = 0.2 kg	g
10	£23·00 ÷ 5	£
11	0.7 litres − 1/2 litre = ☐ mℓ	mℓ
12	3/10 + 2/5	

B

		ANSWER
1	What number is 32 greater than 290?	
2	Write as a decimal 5 tens plus 18 tenths.	
3	How many FIVES must be taken from 3 FIFTIES to leave £1·15?	FIVES
4	How many eighths are there in $7\frac{5}{8}$?	/8
5	29th June is on a Friday. On which day is the 4th July?	
6	Share 75p equally among 8 children. Find (a) how much each (a)	p
	(b) how many pennies are left. (b)	p
7	What mass in kg is double 3 kg 750 g?	kg
8	Which of these numbers will divide exactly by both 6 and 9 without a remainder? 24 36 48 63	
9	Find the area of a playground 30 m long and 18 m wide.	
10	Find the cost of 400 g at 25p per kg.	p
11	From $1\frac{3}{8}$ subtract $\frac{1}{2} + \frac{3}{4}$.	
12	How many degrees in (a) ∠ BDA (a)	°
	(b) ∠ BAC? (b)	°

C

		ANSWER
1	Approximate (a) 9.82 to the nearest whole one (a)	
	(b) £10·48 to the nearest £1 (b)	£
	(c) 3.25 kg to the nearest kg. (c)	kg
2	The kilometre reading on the instrument in a car is 9946.2. What distance has the car to travel for it to read ten thousand kilometres?	km
3	What fraction in its lowest terms is equal to (a) 8 out of 20 (b) 25 out of 40 (c) 70 out of 100? (a) (b) (c)	
4	10 articles cost £2·40. Find the cost of 3.	p
5	Josh was born on 30.6.'03. Write his age in years and months on 1st September 2015.	years months
6	Find the sum of the numbers between 60 and 80 which are divisible by 9.	
7	Find (a) the perimeter of the shape (a)	
	(b) its area. (b)	
8	1000 screws have a mass of 4.2 kg. Find the mass in g of (a) 100 screws (a)	g
	(b) 1 screw. (b)	g
9	A shopkeeper bought 6 balls for £1·32 and sold them to make a total profit of 48p. For how much did he sell each ball?	p
10	A car uses 7 litres of petrol to travel 100 km. How many litres are required for 1600 km?	ℓ
11	Three lines measure 0.04 m; 47 mm; 3.8 cm. Find the difference between the longest and shortest lines.	mm
12	48 centimetre cubes fit exactly into the bottom of this box. The box is 5 cm deep. How many cm cubes are needed to fill it?	

Next work Progress Test 1 on page 16.
Enter the result and the date on the chart.

PROGRESS TEST 1

Write the numbers 1 to 20 down the side of a sheet of paper.
Write alongside these numbers the **answers only** to the following questions.
Work as quickly as you can. Time allowed – **10 minutes.**

1. Write in figures to the nearest hundred six thousand four hundred and fifty.

2. Find the missing number of FIVES.
£1·65 = 2 FIFTIES, 2 TWENTIES, ▢ FIVES.

3. In this set of the factors of 30, one is missing. Which is it?
F = {1, 10, 15, ▢, 6, 30, 3, 2}

4. Write as a decimal the sum of 3 hundreds and 109 hundredths.

5. How many h and min from 11.52 a.m. to 2.27 p.m.?

6. By counting in the given order find the total value of the coins in the box.

7. Find the total when 1.05 is added together 8 times.

8.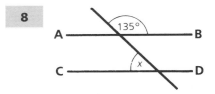

   The lines AB and CD are parallel.
   Find the size of the angle marked x.

9. $\frac{3}{4}$ of a sum of money is 54p. Find the whole amount.

10. 26 × 8 = 208. Write the answer to 260 × 0.8.

11. On a map a road is shown as 9 cm long, which represents a distance of 4.5 km.
Write and complete the scale 1 cm to ▢ m.

12. Take 650 mℓ from 5 litres and give the answer to the nearest 0.5 ℓ.

13. A regular hexagon has sides each measuring 58 mm. Find its perimeter in cm.

14. 200 g of mushrooms cost 48p. Find the price of the mushrooms per $\frac{1}{2}$ kg.

15. The circumference of a wheel measures 2·5 m.
How many times will it turn in travelling 500 m?

16. A garden plot measures 12.8 m long and 8 m wide. Find its area in m².

17. Buttons are bought at 4p each and sold at 7p each.
How much profit is made after selling 200 buttons?

18. The mass of a 10p coin is 6.5 g.
Find the mass in kg of the coins in a £10 bag of TENS.

19. How many 6 cm square tiles are required to cover a rectangular surface measuring 54 cm long and 48 cm wide?

20. Of these 20 examples 6 were wrongly answered.
What fraction, in its lowest terms, was correct?

PROGRESS TEST 1 — RESULTS CHART

You will work Progress Test 1 at **four** different times. When you first work the test
  (a) colour the first column to show the number of examples correct out of 20
  (b) enter the date.
Each time you work the test, enter the result and the date in the marked columns.

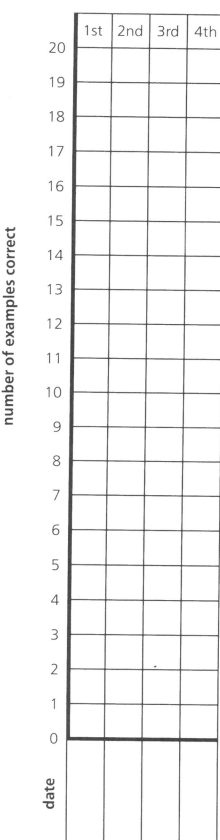

Section 2 Test 1

A   ANSWER

1. Write in figures thirty thousand and fifteen. _____
2. 2006 − 600 _____
3. 3 TENS and 6 FIVES = ▢ TWOS _____ TWOS
4. 10.04 = ▢ hundredths _____ hundredths
5. 56 × 20 _____
6. 36 cm = ▢ m _____ m
7. 0.850 kg = ▢ g _____ g
8. 4.05 + 2.05 _____
9. 0.72 ÷ 8 _____
10. 29 + 13 = 7 × ▢ _____
11. £0·95 = 3 TWENTIES + ▢ FIVES _____ FIVES
12. $\frac{9}{100}$ of £3·00 _____ p

B   ANSWER

1. Find the product of 0.55 and 6. _____
2. What fraction of the square is
   (a) shaded
   (b) unshaded?
   (a) _____ (b) _____
3. Find the total of 79p, 41p and 85p. £ _____
4. Write in 24-hour clock times
   (a) 5 min to 9 in the morning (a) _____
   (b) 10 min past 10 in the evening. (b) _____
5. By how many m is 2 km less than 2.350 km? _____ m
6. What fraction in its lowest terms is 10 out of 60? _____
7. What is the cost of 2 m 10 cm at 40p per m? _____ p
8. By how many kg is $1\frac{1}{2}$ kg greater than 350 g? _____ kg
9. Find the total cost of 100 articles at 7p each. £ _____
10. What is the value in mℓ of the figure underlined in 7.360 litres? _____ mℓ
11. By how much is the total of 7 TENS and 3 FIVES less than £1? _____ p
12. Write in figures to the nearest 100 twenty thousand nine hundred and six. _____

C   ANSWER

1. How many packets each containing 100 cards can be made from thirteen thousand cards? _____
2. Find the change from £1 after paying for 3 oranges at 29p each. _____ p
3. $\frac{1}{7}$ of the mass of a container is 600 g. Find its total mass in kg. _____ kg
4. ABC is an isosceles triangle. Find the size in degrees of the
   (a) angle at B (a) _____ °
   (b) angle at C (b) _____ °
5. Three bottles contain 1.3 litres, 0.9 litres and 0.5 litres. Find in mℓ the average of these quantities. _____ mℓ
6. A bus journey takes 47 min. If a bus starts at 09.50, at what time does it arrive? _____
7. Find the width of the garden plot. (GARDEN PLOT AREA 480 m², 40 m) _____
8. The population of a town is 59 609. Write this number to the nearest 1000. _____
9. The bus fare for a child is half that of an adult. Find the total fares for mother, father and 4 children if a full fare is 54p. £ _____
10. A plan is drawn to a scale of 1 mm to 50 m. What length in km does a line measuring 5 cm on the plan represent? _____ km
11. 40 tiles each 10 cm square cover a surface exactly. What is the area of the surface? _____
12. (SQUARE 8 cm; RECTANGLE 9 cm × 6 cm)
    Which shape has the greater perimeter and by how many cm? _____ by _____ cm

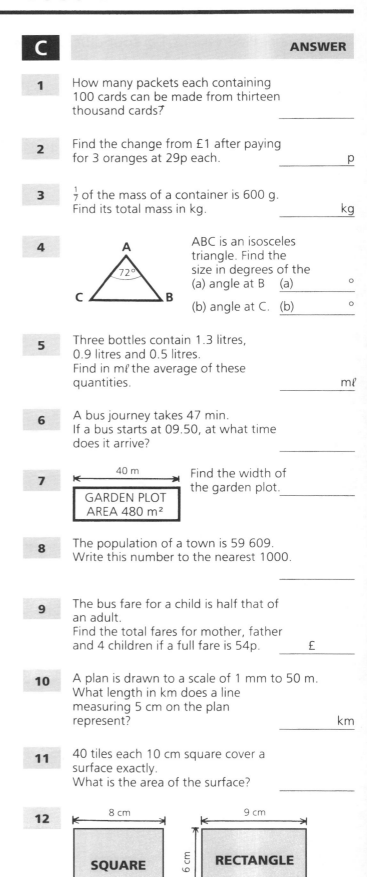

Section 2 Test 2

A

1. Write in words as a decimal the number shown on the abacus picture. (U t h th)
2. 100 × ☐ = 10 000
3. 2.8 km − 300 m = ☐ km
4. 74p × 8 = £☐
5. 3.405 + 0.25
6. 0.065 litres = ☐ mℓ
7. 1/6 of £4·80
8. 47 + 53 + 4000
9. 0.75 min = ☐ s
10. 42 − 15 = ☐ × 3
11. 3 kg + 90 g = ☐ kg
12. 5 × 2 = 0.1 × ☐

B

1. Increase 240 by 1/3 of 60.
2. 13 out of 25 = ☐ out of 100.
3. Find the change from £3·00 after spending £2·19.
4. From twelve thousand take nine hundred and forty.
5. 6 times 85 cm = ☐ m
6. What fraction of the rectangle is (a) shaded (b) unshaded?
7. How many mm are there in 3.070 m?
8. How many times is 250 g contained in 5½ kg?
9. Reduce 70/100 to a fraction in its lowest terms.
10. What is the cost of 100 g at 85p per ½ kg?
11. A square has sides 30 cm long. Find (a) its perimeter (b) its area.
12. Share £100 equally among 8 people. How much each?

C

1. Divide 1655 by 100 and write the answer to the nearest whole one.
2. Of the dots, what fraction in its lowest terms is (a) white (b) coloured?
3. Plums are priced at 3 for 35p. How many can be bought for £2·10?
4. The mass of a box of nails is 3.5 kg. The mass of the box is 600 g. Find the mass of the nails in kg.
5. Mrs Singh lives 20.3 km from her place of work. How many km does she travel in a five-day week making a return journey each day?
6. James' date of birth is 30.8.'94. What will be his age in years on 1st September 2020?
7. Find in cm the length of a diameter which is twice that of the given circle. (35 mm)
8. For how many days would a 200-mℓ bottle of medicine last if 2 spoonfuls were taken 4 times a day? (A medicine spoon holds 5 mℓ.)
9. 15 × 17 = 255 Write the answers to (a) 150 × 170 (b) 1.5 × 17.
10. Find the cost at £5·70 per metre of (a) 10 cm (b) 30 cm.
11. A road on a map measured 4.5 cm which represented a distance of 45 km. What distance does 1 mm on the map represent?
12. (16 cm, 9 cm, 125°, 125°, A) (a) Name the shape. (b) Find its perimeter. (c) Find ∠ A.

Section 2 Test 3

A

		ANSWER
1	2083 − 80	
2	$6 + \frac{7}{10} + \frac{5}{1000}$. Write the answer as a decimal.	
3	2050 mm = ☐ m	_____ m
4	3 h − 25 min = ☐ min	_____ min
5	6 TWOS + 5 FIVES + 3 TENS	_____ p
6	78.5 ÷ 100	
7	$\frac{4}{5} = \frac{☐}{100}$	$\frac{}{100}$
8	0.5 litre − 345 mℓ = ☐ mℓ	_____ mℓ
9	27 + 18 = ☐ × 5	
10	1.8 kg ÷ 3 = ☐ g	_____ g
11	$\frac{2}{3}$ of £3·60	£
12	(clock face with angle A shown) ∠ A = ☐°	_____ °

B

		ANSWER
1	Write as a decimal the total of eleven plus twenty-six thousandths.	
2	Find a number which when multiplied by itself gives as the answer (a) 64 (b) 49.	(a) _____ (b) _____
3	What is left after taking 17p from 4 TWENTIES?	_____ p
4	Write the decimal fraction which is equal to $\frac{3}{20}$.	
5	How many min from 10.55 a.m. to 12.30 p.m.?	_____ min
6	By how many m is 1.4 km longer than 1250 m?	_____ m
7	Find the difference between the largest and smallest of these decimals. 3.4 3.401 3.41	
8	How many pennies remain when 68p is divided by 7?	
9	Multiply 4.050 kg by 100.	_____ kg
10	Find the average of these lengths. 5.4 cm, 4.6 cm, 3.5 cm	_____ cm
11	What speed in km/h is the same as $4\frac{1}{2}$ km in 12 min?	_____ km/h
12	How many times is 260.6 greater than 2.606?	

C

		ANSWER
1	Increase forty thousand by six hundred and three. Write the answer in figures.	
2	Multiply the product of 8 and 7 by 0.1.	
3	On a Celsius thermometer, at what temperature does (a) water boil (b) water freeze?	(a) _____ ° (b) _____ °
4	A car travels 120 km in $1\frac{1}{2}$ h. Find its speed in km/h.	_____ km/h
5	A box contains 10 tins of fruit each having a mass of 223 g. Find the total mass to the nearest $\frac{1}{2}$ kg allowing 230 g for the box.	_____ kg
6	By how many hundredths is 0.82 less than $\frac{17}{20}$?	_____ hundredths
7	(Table: 250g / 100g ; 62p / 25p) The prices of two sizes of bags of bird seed are given. How much is saved by buying the larger bags when purchasing 500 g?	_____ p
8	A line is drawn to a scale of 1 cm to 1 m. (a) Write the scale as a fraction in its lowest terms. (b) What distance in m is represented by 1 mm?	(a) _____ (b) _____ m
9	Which two of these fractions are each equal to $\frac{5}{6}$? $\frac{15}{18}$ $\frac{16}{20}$ $\frac{10}{15}$ $\frac{20}{24}$	_____ _____
10	A ship sails East from Port and then turns SE. At what angle from N is the ship now sailing?	_____ °
11	A rectangle measures 8 cm by 6 cm. What is the length of the sides of a square which has the same perimeter?	_____ cm
12	(Diagram: rectangle 14 cm by 8 cm with shaded triangle) Find these measurements of the shaded triangle. (a) the base (b) the height (c) the area	(a) _____ (b) _____ (c) _____

20

Section 2 Test 4

A

		ANSWER
1	$\frac{1}{4}$ of ten thousand	
2	6 × 0 × 9 × 3	
3	0.9 + 2.145	
4	96p × 6	£
5	1632 ÷ 4	
6	£5·00 − 8 FIVES = £	£
7	$\frac{35}{100} = \frac{\ }{20}$	$\overline{20}$
8	400 g + ▢ g = 0.75 kg	g
9	2 h 25 min + 50 min	h min
10	3.5 km − 2900 m = ▢ m	m
11	£0.77 = 6 TWOS + ▢ FIVES	FIVES
12	[diagram: crossed lines with 82°, angles x and y] Angle x = ▢ ° Angle y = ▢ °	° °

B

		ANSWER
1	Find the product of 0.4 and 9.	
2	What is the answer in g when 24.5 kg is divided by 100?	g
3	From 50 take 0.01.	
4	How many days are there between Christmas Day and New Year's Day?	
5	Find the total of 6 TWENTIES and 7 TENS.	£
6	Write 0.05 as a fraction in its lowest terms.	
7	Which of these numbers will divide into 81 without a remainder? 2 3 5 6 9	
8	Find the cost of 3.5 litres at 26p per $\frac{1}{2}$ litre.	£
9	Which of these decimal fractions is equal to $\frac{3}{4}$? 0.7 0.65 0.75	
10	The diameter of a circle is 15.6 cm. Find its radius in mm.	mm
11	5.4 ÷ 6 = 0.9 Write the answer to (a) 0.54 ÷ 6 (b) 0.054 ÷ 6.	(a) (b)
12	Two angles in a triangle added together make 124°. Find the third angle.	°

C

		ANSWER
1	Find the number which is equal to $(9 \times 10^2) + 6$.	
2	How many 4p sweets can be bought for £6·00?	
3	Find in m the length of ribbon required to make 200 pieces each 8.3 cm long.	m
4	What fraction in its lowest terms is equal to (a) 0.6 (b) 0.16?	(a) (b)
5	Midday Celsius temperatures on three consecutive days are 17°, 15° and 19°. Find the average temperature.	°
6	1 litre of water has a mass of 1 kg. Find the volume in litres of the water in a bottle if the water has a mass of 780 g.	ℓ
7	An article priced at £4·20 was sold in a sale at a reduction of $\frac{1}{3}$. Find the sale price.	£
8	[pin people chart] On the chart each pin person represents 50 people employed in a factory. How many work in the factory?	
9	Jack found the mass of 10 screws to be 95 g. Find the mass in kg of 100 screws.	kg
10	[equilateral triangle ABC, 60° at A, 86 mm base CB] ABC is an equilateral triangle. Find (a) its perimeter in cm (b) the size of the angles at B and C.	cm °
11	Yasmin went shopping with a £5 note in her purse. She had left 2 FIFTIES, 1 TEN and 6 TWOS. How much had she spent?	£
12	[rectangle 12 cm × 9 cm with triangle inside, base 7 cm] Find the area of (a) the rectangle (b) the triangle.	

Turn back to page 16 and work for the second time Progress Test 1.

Enter the result and the date on the chart.

Section 2 Test 5

A

		ANSWER
1	20 000 = 20 × 10 × ☐	
2	0.817 = 8 tenths + ☐ thousandths	_____ thousandths
3	57 × 70	
4	3.2 m ÷ 8 = ☐ cm	cm
5	24 − (18 × 0)	
6	£20·36 = ☐ TENS + 6p	TENS
7	$\frac{4}{5} - \frac{1}{2}$	
8	65 mℓ × 100 = ☐ litres	ℓ
9	2.6 cm + 3.9 cm = ☐ mm	mm
10	£ ☐ × 9 = £40·50	£
11	1.5 kg − 280 g = ☐ g	g
12	6 × 5 = 0.1 × ☐	

B

		ANSWER
1	Write as a decimal 1035 thousandths.	
2	Approximate 59.7p to the nearest penny.	p
3	(a) $\frac{9}{10} = \frac{☐}{100}$	(a) ____/100
	(b) Write the fraction as a decimal.	(b)
4	Which two of these angles when added together make two right angles? 67°, 103°, 87°, 113°	____° ____°
5	How many times is 0.48 less than 480?	
6	Find in mm the value of the figure underlined. 7.0_7_5 m	mm
7	9 out of 25 = ☐ out of 100.	
8	How much change out of a £5 note after spending £1·46?	£
9	$\frac{1}{2}$ kg of tomatoes costs 80p. Find the cost of 100 g.	p
10	What fraction of $\frac{1}{4}$ ℓ is 150 mℓ?	
11	Which of these fractions is between one half and one quarter in size? $\frac{3}{5}$ $\frac{2}{3}$ $\frac{7}{10}$ $\frac{3}{8}$	
12	Find the area of (a) the rectangle (b) the shaded triangle.	(a) _____ (b) _____

(Rectangle 10 cm × 7.4 cm with shaded triangle)

C

		ANSWER
1	How many bottles each holding $\frac{1}{4}$ litre can be filled from $7\frac{1}{2}$ litres?	
2	A packet of 100 sheets of paper costs £2·30. Find the cost of 150 sheets.	£
3	Take the least of these decimal fractions from the greatest. 0.84 0.9 0.865 0.897	
4	Josh swims 15 lengths of the bath which is 20 m long. How many m short of $\frac{1}{2}$ km does he swim?	m
5	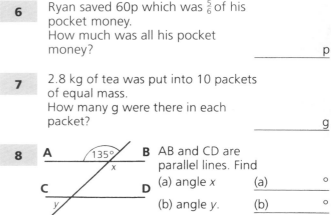 The large cube is made from a number of centimetre cubes. (a) How many cm cubes are there? (b) 4^3 = ☐	(a) _____ (b) _____
6	Ryan saved 60p which was $\frac{5}{6}$ of his pocket money. How much was all his pocket money?	p
7	2.8 kg of tea was put into 10 packets of equal mass. How many g were there in each packet?	g
8	AB and CD are parallel lines. Find (a) angle x (b) angle y.	(a) ____° (b) ____°
9	The total cost of 2 full fares and 1 half fare was £55. Find (a) the full fare (b) the half fare.	(a) £ ____ (b) £ ____
10	What is the area in m² of a path which measures 29 m long and 50 cm wide?	m²
11	A distance of 9.875 km was written as '10 km to the nearest km'. By how many m was the approximation incorrect?	m
12	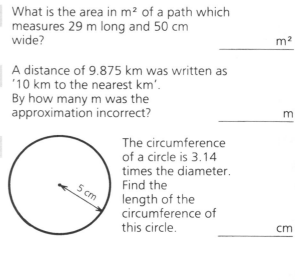 The circumference of a circle is 3.14 times the diameter. Find the length of the circumference of this circle.	cm

Section 2 Test 6

A

1. 1200 ÷ 30
2. 7p + 5p + 8p + 25p ___ p
3. (7 × 9) − (8 + 5)
4. $\frac{6}{25} = \frac{\ }{100}$ ___/100
5. 900 g + 500 g = ___ kg
6. $4.908 = 4 + \frac{9}{10} + \frac{\ }{1000}$ ___/1000
7. £0·86 − 19p = ___ p
8. 2 km − 540 m = ___ m
9. 9.92 + 0.02
10. 250 mℓ × 6 = ___ litres ___ ℓ
11. 2 m 80 cm ÷ 4 = ___ cm
12. 180° − (63° + 57°) ___°

B

1. Write as a decimal the total of 200, 6, $\frac{7}{10}$ and $\frac{9}{1000}$.
2. From the product of 7 and 9 take their sum.
3. Write as a vulgar fraction in its lowest terms (a) 9 out of 30 (b) 24 out of 40. (a) ___ (b) ___
4. How many days altogether are there in the months of April and May?
5. Find in g the value of the figures underlined. 4.07<u>5</u> kg ___ g
6. What must be added to 9.405 to make 10?
7. Change to decimal fractions (a) $\frac{27}{50}$ (b) $\frac{9}{20}$.
8. If ∠A = ∠B find the size of each angle. (130°) ___°
9. Divide £2·75 by 10 and write the answer to the nearest penny. ___ p
10. Find the cost of 1.2 m of tape at 40p per m. ___ p
11. Take 6 thousandths from 8 point 4.
12. Find in cm the perimeter of an equilateral triangle, the sides of which measure 27 mm. ___ cm

C

1. By how many is 8^2 greater than 3^3?
2. Approximate 25 948 (a) to the nearest 1000 (b) to the nearest 100.
3. Find the date which is 9 days after 26th September.
4. How many TWENTIES are there in ten £10 bags of TWENTIES?
5. What is the difference in mℓ between the largest and smallest of these quantities? $\frac{1}{2}$ litre, 450 mℓ, $\frac{3}{5}$ litre, 620 mℓ ___ mℓ
6. ABC is a right-angled triangle. Find in degrees (a) the angle marked x (b) the angle marked y.
7. A length of plastic strip 54 cm long is cut into two pieces so that one is five times as long as the other. What is the length of each piece? ___ cm ___ cm
8. George spent £80 which was $\frac{2}{5}$ of his savings. Find the amount of his savings at first. £ ___
9. On a map, a line 3 cm long represents 150 km. Find the scale to which the map is drawn by completing 1 mm to ___ km.
10. A 5p coin has a mass of 3.25 g. Find in g the mass of 40 FIVES. ___ g
11. A circular tin has a diameter of 6.4 cm. What is (a) the length, (b) the width of the smallest rectangular tray which would contain a single row of 8 tins? (a) ___ cm (b) ___ cm
12. Find the area of (a) the triangle ABC (b) a triangle having the same base but half the height of ABC.

Section 2 Test 7

A

1. Write in figures the number forty thousand six hundred and five.
2. (55 − 8) − (7 × 0)
3. 12 weeks = ☐ days
4. $\frac{3}{10}$ of $1\frac{1}{2}$ litres = ☐ ml
5. Write as a decimal fraction (a) $\frac{59}{100}$ (b) $\frac{7}{100}$.
6. 10.2 − 3.07
7. 5.405 m = ☐ mm
8. 4 TWENTIES + 2 TENS + 9 TWOS = £ ☐
9. (a) 0.45 = $\frac{☐}{100}$ (b) 0.08 = $\frac{☐}{25}$
10. 208 g + 1.5 kg = ☐ kg
11. ☐ p × 6 = £4·56
12. $\frac{755}{7}$ = ☐ rem. ☐

B

1. (a) What decimal fraction of the 100 small squares is shaded?
   (b) Write this decimal fraction as a vulgar fraction.
2. 47 out of 100 = 47 per cent (%). Write as a % (a) 79 out of 100 (b) 4 out of 100.
3. By how much is £15 more than 15 FIFTIES?
4. Find in ml the value of the figures underlined. 8.305 litres
5. Express in km/h a speed of 90 km in 45 min.
6. How many days are there between 28th November and 7th December?
7. 89% of a sum of money is spent. What percentage is left?
8. Find the cost of 600 g at £0·50 per $\frac{1}{2}$ kg.
9. Name the parallel sides in the parallelogram. (PARALLELOGRAM ABCD)
10. Approximate (a) 10 046 to the nearest 100 (b) 3 kg 370 g to the nearest 0.5 kg.
11. Find the change from a £10 note after spending £7·62.
12. The area of a rectangle is 47.5 cm². Its width is 5 cm. Find its length.

C

1. Write each of these fractions with a denominator of 100. (a) $\frac{3}{10}$ (b) $\frac{2}{5}$ (c) $\frac{1}{25}$
2. A school was built in 1902. For how many years will it have been in use by the year 2020.
3. What decimal fraction is equal to (a) 57% (b) 8%?
4. How many packets each containing 365 g can be made from $36\frac{1}{2}$ kg?
5. Write the next two decimal numbers in this series. 1.75, 2.0, 2.25, ☐, ☐
6. Prices in a shop were increased by $\frac{1}{3}$. Find the new price of articles which cost (a) 48p (b) 84p.
7. From a can holding 10 litres of oil, 2.75 litres are drawn off. Find in litres and ml the quantity which remains.
8. The sides of the cube each measure 5 cm. Find the area of (a) 1 face (b) all the faces.
9. The average length of 3 pieces of wood is 13 cm. Two of the pieces measure 12 cm and 18 cm. What is the length of the third piece?
10. 10 articles cost £8·40. Find the cost of 3 articles.
11. Mortar is made by mixing 1 part cement with 4 parts sand. Find the mass of (a) cement (b) sand to make 30 kg of mortar.
12. 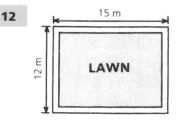 The path round the lawn is 1 m wide. Find the area of the lawn.

Section 2   Test 8

A

1. 50 000 = 500 × ☐
2. (27 ÷ 3) = (☐ ÷ 7)
3. (a) $\frac{1}{2}$ = ☐ out of 100
   $\frac{1}{2}$ = ☐ %
   (b) $\frac{1}{4}$ = ☐ out of 100
   $\frac{1}{4}$ = ☐ %
4. $1\frac{1}{2} - \frac{7}{10}$
5. $5\frac{1}{2}$ min = ☐ s
6. 18.3 + 18.3 + 18.3 + 18.3 + 18.3
7. £0·19 × 6
8. $\frac{17}{20} = \frac{☐}{100} = ☐ \%$
9. 4.385 − 4.325
10. $\frac{2}{3}$ of 75p
11. 850 g × 4 = ☐ kg
12. £4·20 = 6 FIFTIES + ☐ TWENTIES

B

1. Which of these numbers are multiples of 8?
   36   48   60   72
2. Write as a percentage (a) $\frac{63}{100}$
   (b) 0.07.
3. What distance is travelled in 15 min at 72 km/h?
4. Find the cost of 750 g at 48p per kg.
5. What must be added to £4·74 to make £6·00?
6. Write in 24-hour clock time a quarter of an hour earlier than 23.10.
7. Find the area of the right-angled triangle. (6 cm, 10.5 cm)
8. In a class 36% of the children were boys. What percentage were girls?
9. 360° − (80° + 75° + 120°) = ☐ °
10. Find to the nearest penny $\frac{1}{8}$ of 95p.
11. Find the average of these volumes. 650 mℓ, 800 mℓ, 350 mℓ
12. What is the perimeter in cm of a regular HEXAGON each side of which measures 58 mm?

C

1. Write these decimal fractions as percentages. (a) 0.3
   (b) 0.07
   (c) 0.84
2. Leah had in her purse 3 FIFTIES, 2 TWENTIES and 2 FIVES. She spent £1·17. How much had she left?
3. A girl wrote 7 tens instead of 7 tenths for an answer. By how much was her answer wrong?
4. What is the value of 1% of
   (a) £1
   (b) 1 kg?
5. The diagram represents a length of 100 m. Find the length represented by the shaded part.
6. Shahid spends 38% of his pocket money on bus fares and 24% on sweets. What percentage of his money remains?
7. A penny has a mass of 3.56 g. Find the mass in kg of £10 of pennies.
8. The radius of the inner circle is 56 mm and the shaded ring is 16 mm wide. Find in cm the diameter of the outer circle.
9. 20 tins of sweetcorn were bought for £10.80 and sold at 60p per tin. Find the profit after selling all the tins.
10. "Round off" each of these numbers to the nearest whole one.   (a) $9\frac{7}{12}$
    (b) 100.06
    (c) 4.815
11. Tom received 2p each time Megan received 3p. How much did they each receive from a total of £2·00?
    Tom ☐ p   Megan £ ☐
12. How many tiles each 20 cm square are needed to cover the surface? (5 m × 1 m)

Turn back to page 16 and work for the third time Progress Test 1. Enter the result and the date on the chart.

25

Section 2   Test 9

A

1. Write in figures one hundred and four thousand.
2. £1·11 + 8p + 9p + 24p
3. 0.35 = ▢ thousandths
4. 2.8 litres − 630 ml
5. 350 g × 5 = ▢ kg
6. $\frac{57}{100}$ of £500
7. 7 m 28 cm ÷ 8 = ▢ cm
8. 297 + 103 = 40 × ▢
9. $1 - (\frac{3}{10} + \frac{2}{5})$
10. 8 × 7 = ▢ × 0.1
11. £ $\frac{14·42}{7}$ = £ ▢
12. (a) $\frac{1}{4}$ = 0.25 = ▢ %
    (b) $\frac{3}{4}$ = 0.75 = ▢ %

B

1. Write as decimal fractions (a) $\frac{3}{20}$ (b) $\frac{8}{25}$.
2. What is the difference in g between $1\frac{3}{4}$ kg and 950 g?
3. What length when divided by 6 is equal to 2 m 50 cm?
4. The three angles of a triangle each measure 60°. Name the triangle according to (a) its angles (b) its sides.
5. Find the value of
   (a) 50% of 650
   (b) 25% of £0·72.
6. How many times is 200 ml contained in 2.8 litres?
7. Approximate to the nearest cm
   (a) 20.3 cm
   (b) 79 mm.
8. Write in its lowest terms the vulgar fraction which is equal to 75%.
9. Find the length in m which is equal to the figures underlined. 6.5<u>35</u> km
10. At 75p per $\frac{1}{2}$ kg find the cost of
    (a) 100 g
    (b) 300 g.
11. Find in cm the perimeter of a rectangle which is 65 mm long and 30 mm wide.
12. (a) 0.1 = ▢ %
    (b) 0.4 = ▢ %

C

1. 18% of the people at a concert were men, 39% were women and the rest were children. What percentage were children?
2. Katie went on holiday on 29th July and returned on 7th August. For how many days was she on holiday?
3. Josh saved $\frac{1}{4}$ of his pocket money and spent $\frac{1}{2}$ of the remainder. What fraction did he spend?
4. The population of a small town was ten thousand. 30% of the people were under the age of 25. How many was that?
5. What is the length in cm of the longest straight line which can be drawn in this circle? (54 mm)
6. 50% of a sum of money was £7·00. Find (a) the whole sum of money
   (b) 75% of all the money.
7. How many packets each containing 0.2 kg can be made from 35 kg?
8. A fruit drink contains 7 parts of water and 1 part of juice. How many ml of each are required to make 2 litres?
9. A boy's walking pace measures 60 cm. How many m has he walked after taking 50 paces?
10. The mass of 120 kg of dry sand is increased when wet by 10%. Find its mass when wet.
11. An article costing £25 is reduced by $\frac{1}{20}$ for cash payment. Find
    (a) the price reduction
    (b) the cash payment price.
12. Find the area of
    (a) the front
    (b) the end
    (c) the bottom of the box.

(Box: 16 cm × 7 cm × 4.5 cm)

Section 2 Test 10

A

1. 6009 + ☐ = 8000
2. £3·75 + £3·75 + £3·75 + £3·75   £ _____
3. $\frac{7}{20}$ of £2·00   _____ p
4. $16\frac{2}{3}$ = ☐ thirds $\frac{\_\_\_}{3}$
5. 19 min + ☐ h ☐ min = 3 h   ___ h ___ min
6. $\frac{11}{25} = \frac{☐}{100}$ = ☐ % $\frac{\_\_\_}{100}$ = ___ %
7. £2·00 − (78p + 30p) = ☐ p   _____ p
8. $\frac{3}{10}$ of 1 m 80 cm = ☐ cm   _____ cm
9. 0.48 × 9
10. $\frac{2 \text{ kg } 400 \text{ g}}{5}$ = ☐ g   _____ g
11. (a) 0.7 = ☐ %   (a) ___ %
    (b) 0.9 = ☐ %   (b) ___ %
12. £0·64 × 6 = £   £ _____

B

1. Write as a fraction, in its lowest terms, and then as a percentage, the part of the large square which is
   (a) shaded   (a) ___ %
   (b) unshaded.   (b) ___ %
2. How many right angles are there in 270°?
3. By what length is 50.4 km longer than $47\frac{1}{2}$ km?   ___ km
4. Increase £3·70 by 10%.   £ ___
5. How many ml must be added to 3050 ml to make $3\frac{1}{4}$ litres?   ___ ml
6. $\frac{1}{4}$ of a sum of money is 79p. What is 50% of the money?   £ ___
7. Express in km/h a speed of 6.5 km in 10 min.   ___ km/h
8. Write as a fraction in its lowest terms
   (a) 20%   (a) ___
   (b) 2%.   (b) ___
9. Approximate to the nearest $\frac{1}{2}$ kg
   (a) 7.650 kg   (a) ___ kg
   (b) 3850 g.   (b) ___ kg
10. Find the product of 0.2 and 0.3.
11. (a) Write 20% as a decimal fraction.   (a) ___
    (b) What is 20% of one thousand?   (b) ___
12. Use the formula A = lb to find the area of a rectangle when l = 16.5 cm and b = 9 cm.   ___ cm²

C

1. Find the profit on an article which was bought for £8·35 and sold for £10.   £ ___
2. Two angles of a triangle each measure 78°.
   (a) Name the triangle by its sides.   (a) ___
   (b) Find the third angle.   (b) ___ °
3. |←──────── 7.8 cm ────────→|
   The line has been drawn to a scale 1 mm to 10 m.
   What distance in m does it represent?   ___ m
4. 40% of a sum of money is £36. Find
   (a) 10% of the sum of money   £ ___
   (b) the whole sum of money.   £ ___
5. (a) Name the regular shape drawn in the circle.   ___
   (b) What is the size of the angle shaded at the centre?   ___ °
6. A bus runs every 35 min starting at 07.40.
   At what time does the third bus leave?
7. How many badges at 3 for 20p can be bought for £3·80?
8. The circumference of a circle is found from the formula C = πd.
   Find the circumference when π = 3.14 and d = 5 cm.   ___ cm
9. Use the formula $A = \frac{bh}{2}$ to find the area of the triangle.   ___ cm²
10. 0.5 kg of mushrooms cost £1·20.
    What is the cost of (a) 100 g   (a) £ ___
                        (b) 0.9 kg?   (b) £ ___
11. This shape is built from cm cubes. Find its dimentions.
    (a) length   (a) ___ cm
    (b) breadth   (b) ___ cm
    (c) height   (c) ___ cm
12. How many cm cubes are used to build the block?

Section 2 Test 11

A

1. $10^3 = 10 \times 10 \times 10 = \square$
2. $1100 - 280$
3. $490\,g = \square\,kg$
4. $3.094 + 0.06$
5. (a) $\frac{9}{100} = \square\,\%$
   (b) $0.36 = \square\,\%$
6. $3\frac{3}{4} - 2\frac{7}{8}$
7. $9\text{ metres} \div 30 = \square\,cm$
8. $£0.08 \times 14 = £\square$
9. $60\% = \frac{\square}{5}$
10. $175\text{ seconds} = \square\text{ min }\square\text{ s}$
11. $0.073 = \square\text{ thousandths}$
12. 75% of $1.2\,kg = \square\,g$

B

1. Find the total in litres of 700 mℓ, 300 mℓ and 450 mℓ.
2. How much greater than $\frac{1}{2}$ is
   (a) 0.64
   (b) 0.502?
3. What is the time $\frac{3}{4}$ h later than 23.20?
4. Find the cost of 50 g at £1·40 per $\frac{1}{2}$ kg.
5. How many times is 370 contained in 37 000?
6. Write each of the following as (a) an improper fraction (b) a mixed number.
   67 tenths
   23 sixths
7. Find the difference between $\frac{3}{4}$ of £1 and 0.8 of £1.
8. How many g are there in $\frac{1}{8}$ of 1 kg?
9. Write each of these percentages as (a) a decimal, (b) a fraction in its lowest terms.
   80%
   30%
10. What is the value of x when 39 less than x is 76?
11. What percentage of £1·00 is
    (a) 54p
    (b) 6p?
12. Find in m² the area of a rectangle 16.8 m long and 50 cm wide.

C

1. Find the numbers between 30 and 50 which have both 4 and 6 as factors.
2. The average amount saved by 7 children was 83p. How much was saved altogether?
3. From the diagram find and name three obtuse angles.
4. The distance between two towns is 580 km. If a map is drawn to a scale of 1 cm to 100 km what length in mm represents this distance?
5. Write as a fraction in its lowest terms
   (a) 100 g of 400 g
   (b) 9 km or 45 km
   (c) 500 mℓ of $1\frac{1}{2}$ litres.
6. What is the Celsius temperature for 7° below zero?
7. How long in h and min will it take a car to travel 325 km at a speed of 100 km/h?
8. Find the missing divisor in this example.
   $\square\overline{)87}$ 9 rem. 6
9. A bag contained an equal number of TENS and FIVES to a total value of £4·50. How many coins were there of each kind?
10. By putting in a decimal point make the 7 in each number have the value of 7 thousandths. (a) 2070 (b) 67
11. Find the least number of pence which must be added to £2·55 to make it divisible exactly by 7.
12. A carpet 6.5 m by 4 m is fitted on to a floor to leave a border of 50 cm wide. Find
    (a) the length
    (b) the breadth of the room.

Turn back to page 16 and work for the fourth time Progress Test 1. Enter the result and the date on the chart.

Section 2   Test 12

A

1. 70 000 = 10 × 100 × ▢
2. 4.505 = ▢ thousandths _____ thousandths
3. 3 m 40 cm + 1.8 m = ▢ m ▢ cm _____ m _____ cm
4. 40 × $2\frac{3}{4}$
5. 4.7 kg + 2.4 kg = ▢ g _____ g
6. $\frac{1}{20} = \frac{▢}{100} = ▢\%$ _____/100 = _____ %
7. 20 000 + 800 + 6
8. 5.35 − 0.035
9. £70 ÷ 20   £ _____
10. 40% of 900
11. 350 ml × 8 = ▢ litres _____ ℓ
12. 100% of £3·86   £ _____

B

1. By how many is 30 090 less than forty thousand?
2. Find the total value of these amounts. 47p, 62p, 13p and 18p.   £ _____
3. What fraction of 1 hour is (a) 12 min (b) 50 min?   (a) _____ (b) _____
4. Find in pence 5% of £4·00.   _____ p
5. How many degrees are there turning clockwise
   (a) from SE to W   (a) _____ °
   (b) from N to SW?   (b) _____ °
6. Find the difference between the largest and smallest of these numbers. 2.202, 2.02, 2.22
7. What is the cost of 800 ml at 90p per litre?   _____ p
8. By how many twelfths is $\frac{1}{4}$ less than $\frac{1}{3}$?
9. What percentage is
   (a) 20 of 200   (a) _____ %
   (b) 250 g of 1 kg?   (b) _____ %
10. Find the value of x when $\frac{x}{5} = 3 \times 9$.
11. Find to the nearest penny $\frac{£7·87}{8}$.   _____ p
12. Find the area in cm² of a triangle with a base of 75 mm and a height of 10 cm.   _____ cm²

C

1. What number when divided by 6 equals the product of 5 and 9?
2. How many h and min are there from 09.40 to 13.25?   _____ h _____ min
3. RECTANGLE, RHOMBUS, PARALLELOGRAM, TRAPEZIUM
   Of these quadrilaterals which has
   (a) 4 equal sides   (a) _____
   (b) one pair of parallel sides only?   (b) _____
4. 20% of 700 children at a school stay for dinner. How many
   (a) stay   (a) _____
   (b) go home for dinner?   (b) _____
5. How many sectors of the given size can be cut from the circle? (120°)
6. 7 packets of equal mass together have a mass of 4.550 kg. Find in g the mass of each packet.   _____ g
7. Which of these numbers are multiples of both 6 and 8? 16, 24, 30, 36, 42, 48, 64
8. One part of weed-killer is mixed with 5 parts of water. How many ml of each are required to make 1.5 litres of weed-killer?   _____ ml   _____ ml
9. Jack buys a bicycle for £80 and pays for it by weekly instalments of 5%.
   (a) For how many weeks does he pay?   (a) _____
   (b) How much is the weekly payment?   (b) £ _____
10. The area of a triangle is 72 cm² and its height is 8 cm. Find the length of the base.   _____ cm
11. A car travels 10 km on 1 litre of petrol. How much petrol does the car use if it travels on average 450 km daily for 6 days?   _____ ℓ
12. (a) How many cm cubes fit exactly into the bottom of the box?   (a) _____
    (b) If the box is 3 cm high, how many cm cubes are needed to fill it?   (b) _____
    (7 cm, 15 cm)

Next work Progress Test 2 on page 30.
Enter the result and the date on the chart.

PROGRESS TEST 2

Write the numbers 1 to 20 down the side of a sheet of paper.
Write alongside these numbers the **answers only** to the following questions.
Work as quickly as you can. Time allowed – **10 minutes.**

1. Write in words the number which is equal to $(7 \times 10^3) + (8 \times 10^2)$.

2. Katie has saved £2, 5 TWENTIES, 7 TENS, 6 FIVES and 8 TWOS. By how much is the total less than £10?

3. What fraction in its lowest terms is 150 ml of 750 ml?

4. $\frac{7}{8}$ of a sum of money is 63p. Find the amount of all of the money.

5.  The diagram shows how Mrs Wilson spends her housekeeping money. What percentage of the money is spent on food?

6. $59 \times 37 = 2183$. Write the answer to 0.59×37.

7. Using the scale 1 mm to 0.5 m find the length of a line in cm which is drawn to represent 67.5 m.

8. 90% of the children in a school of 580 went on a school journey. How many children remained at school?

9. 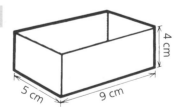 Find to the nearest km the distance from Troup to Ling.

10. The width of a rectangular garden path is 50 cm. Its area is 27.5 m². Find its length.

11. Find the mass in kg and g of 20% of 18.5 kg.

12. How much less than $\frac{1}{2}$ is 0.094?

13. 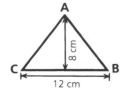 In this parallelogram find in degrees the angle marked *y*.

14. Prices at a sale were reduced by 5%. How much is paid for an article priced at £1·80?

15. How many discs of 4-cm radius can be fitted exactly into one layer at the bottom of a rectangular tray which is 64 cm long and 16 cm wide?

16. Find to the nearest penny the cost of 1.5 m of ribbon at £1·05 per m.

17. How many cm cubes are needed to fill the box?

18. How many times is 30 greater than 0.03?

19. How much cheaper is 300 g of broccoli at 80p per $\frac{1}{2}$ kg than the same mass at £1·10 per $\frac{1}{2}$ kg?

20. By how many cm² is the area of the triangle ABC greater than a triangle on the same base but half the height?

PROGRESS TEST 2 — RESULTS CHART

You will work Progress Test 2 at **four** different times. When you first work the test
   (a) colour the first column to show the number of examples correct out of 20
   (b) enter the date.
Each time you work the test, enter the result and the date in the marked columns.

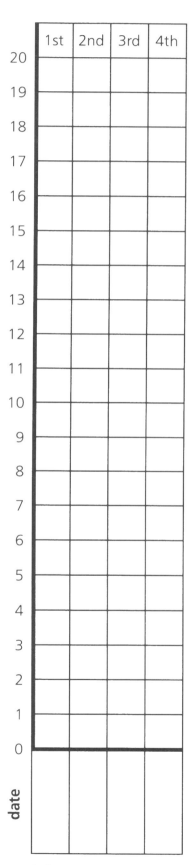

Section 3 Test 1

A

1. Write in figures two hundred and six thousand and forty.
2. 4.008 = ☐ thousandths _____ thousandths
3. 39 × 200
4. 17 TWENTIES and 6 TENS = £ ☐   £ _____
5. 1.48 − 0.8
6. 7.045 km = ☐ km ☐ m   _____ km _____ m
7. 10% of 3 kg = ☐ g   _____ g
8. $\frac{2}{5}$ of £1·40   _____ p
9. 450 mℓ × 8 = ☐ litres   _____ ℓ
10. (a) $\frac{7}{12} = \frac{21}{☐}$ _____ 21
    (b) $\frac{27}{30} = \frac{☐}{10}$ _____ 10
11. $\frac{3}{4}$ h − 18 min = ☐ min   _____ min
12. $\frac{£8·96}{7}$ £ _____

B

1. Find the total of 17p, 34p and 29p. Write the answer as £s.   £ _____
2. What fraction in its lowest terms is 150 mℓ of 1 litre?
3. Approximate to the nearest whole number (a) $9\frac{2}{3}$ (a) _____
   (b) 24.06. (b) _____
4. Find the cost of 5.2 m at 40p per m.   £ _____
5. By how many is 99 060 less than one hundred thousand?
6. Write the 24-hour clock time which is 16 min later than 23.45.
7. How many times is $\frac{3}{4}$ contained in $4\frac{1}{2}$?
8. What percentage is
   (a) 3p of £1·00 (a) _____ %
   (b) 35 cm of 1 m? (b) _____ %
9. Find the difference between 0.850 ℓ and 900 mℓ.   _____ mℓ
10. What mass in kg is 7 times 350 g?   _____ kg
11. Find the perimeter of a rectangle which measures 8.4 cm by 5.9 cm.   _____ cm
12. x × 9 = 7 m 200 mm. Find the length in mm which is equal to x.   _____ mm

C

1. How many hundredths must be added to 3.81 to make a total of 4?   _____ hundredths
2. A bottle holds 250 mℓ of medicine. How many bottles can be filled from 10 litres?
3. Of the people attending a football match 57% were men, 29% were children and the remainder women. What percentage were women?   _____ %
4. A car travels at a speed of 70 km/h. How far does it travel in 90 min?   _____ km
5. By how many pennies is 7 TENS greater than the total of 8 FIVES and 9 TWOS?   _____ p
6. What fraction of the circle is (a) unshaded (a) _____
   (b) shaded? (b) _____
7. A metal strip is 20 cm long. How many such strips can be cut from 2 lengths each 5 m 60 cm?
8. A bus arrived at the station at 18.23 but it was 35 mins late due to fog. Find the correct arrival time.
9. The price for $\frac{1}{2}$ kg of carrots in 4 consecutive weeks was 36p, 32p, 28p, 24p. Find the average price per $\frac{1}{2}$ kg.   _____ p
10. ABC is a triangle inscribed in a semicircle. Find in degrees the size of ∠ BAC   _____ °
    ∠ ABC.   _____ °
11. A map is drawn to a scale of 1 cm to 1 km. Express this scale
    (a) as a fraction (a) _____
    (b) as a ratio. (b) _____
12. Find in cm² the area of
    (a) the front (a) _____ cm²
    (b) the end (b) _____ cm²
    (c) the bottom of the box. (c) _____ cm²

Section 3   Test 2

A

1. Write in words the number 410 006.
2. 375 m = ☐ km _____ km
3. 4.355 + 0.45
4. 85 ml × 100 = ☐ litres _____ l
5. (a) 7% of £1·00   (a) _____ p
   (b) 7% of £6·00   (b) _____ p
6. $\frac{5}{8}$ of 72p   _____ p
7. 0.875 kg = ☐ g   _____ g
8. 108 min = ☐ h ☐ min   _____ h _____ min
9. 634 ÷ 1000
10. £0·08 × 30   £ _____
11. 5.2 − 0.17
12. 25% of 150

B

1. Write as a decimal the total of 10, $\frac{3}{10}$ and $\frac{17}{1000}$.
2. Find the product of 7, 9 and 5.
3. What is the cost of 750 g at 48p per $\frac{1}{2}$ kg?   _____ p
4. By how many sixths is $\frac{2}{3}$ greater than $\frac{1}{2}$?
5. [A | B | C]
   The strip is divided into 3 parts. What percentage of the whole is each part?
   A _____ %
   B _____ %
   C _____ %
6. Write in g the value of the 7 in 6.875 kg.   _____ g
7. Find the number of days, not counting the first, from 21st June to 7th July.
8. Decrease £3·50 by 20%.   £ _____
9. Add together the largest and smallest of these numbers.
   2.01, 2.11, 2.001, 2.101
10. By how many hundredths is 9.07 less than 10?   _____ hundredths
11. Write (a) £10·62 to the nearest £   (a) £ _____
    (b) 3 l 330 ml to the nearest $\frac{1}{2}$ l.   (b) _____ l
12. Find in cm² the area of a rectangle 15 cm long by 50 mm wide.   _____ cm²

C

1. A man cycled 30 km each day for a fortnight. How many km did he cycle altogether?   _____ km
2. Samina paid 46p with a £1 coin and received three coins as change. Name the three coins.   _____ p _____ p _____ p
3. The circumference of a trundle wheel is 1 m. How many times does the wheel turn in going 0.75 km?
4. In the number 28.038 how many times is the 8 marked x smaller than the 8 marked y? (y above first 8, x above second 8)
5. 3 m of wire cost £3·60. Find the cost of 15 m.   £ _____
6. O is the centre of the circle. ∠ OAB is 56°. Find
   (a) ∠ OBA   (a) _____ °
   (b) ∠ AOB.   (b) _____ °
7. A parcel has a mass of 1.8 kg. Find the mass in g of a parcel which is two-thirds of this mass.   _____ g
8. The dates of birth of three children are given.
   16.1.'99
   18.4.'99
   17.7.'98
   (a) By how many months is the oldest child older than the youngest?   (a) _____ months
   (b) In which year will the youngest child be 35 years old?   (b) _____
9. 2.25 litres of milk are poured in equal amounts into 5 glasses. How many ml are there in each glass?   _____ ml
10. A line about which a shape balances is called an axis of symmetry. Which of these shapes A, B, C or D has one axis of symmetry?
11. A boy received a gift of a £10 note. He spent £3. What % did he save?   _____ %
12. Find the area of the shape in cm². (8 cm top, 10 cm right, 15 cm bottom, 6 cm left)   _____ cm²

33

Section 3 Test 3

A

		ANSWER
1	50 × 10 × 1000	
2	1.2 km − 900 m = ▢ m	_____ m
3	48p × 7	£ _____
4	29 + 25 = 6 × ▢	_____
5	$5 - 3\frac{3}{8}$	_____
6	75% of £10	£ _____
7	3 kg ÷ 8 = ▢ g	_____ g
8	(a) 0.45 = ▢ % (b) $\frac{11}{50}$ = ▢ %	(a) _____ % (b) _____ %
9	400 mℓ + 250 mℓ + 500 mℓ = ▢ ℓ	_____ ℓ
10	0.84 = ▢ thousandths	_____ thousandths
11	8 FIVES + ▢ TWOS = 60p	_____ TWOS
12	70° + 38° + ▢° = 180°	_____ °

B

		ANSWER
1	Write in words the number 1 000 000.	
2	How much change from £4·00 after spending £3·26?	_____ p
3	What percentage of 2 kg is 500 g?	_____ %
4	Find the cost of 1 m 30 cm at 80p per m.	£ _____
5	What is the difference in mm between 3.9 cm and 4.6 cm?	_____ mm
6	Divide the sum of 38 and 27 by 5.	
7	Find the reflex angle AOB. (124°)	_____ °
8	What is the time in h and min from 10.15 a.m. to 12.05 p.m.?	_____ h _____ min
9	How many times is 300 mℓ contained in 1.8 litres?	
10	The total mass of 5 parcels is 2 kg 400 g. Find the average mass of the parcels.	_____ g
11	What fraction in its lowest terms is equal to (a) 15% (b) 4%?	(a) _____ (b) _____
12	How many cm² are there in 1 square metre?	_____ cm²

C

		ANSWER
1	Katie wrote the total of £2·50, £3·50 and £2·75 as £9·25. By how much was her total wrong?	_____ p
2	Complete the set of square numbers between 10 and 101 by finding x and y. S = { 16, 25, 36, x, 64, y, 100 }	_____ x _____ y
3	1 litre of water has a mass of 1 kg. Find the mass of 850 mℓ of water.	_____ g
4	A road shown on a map measures 50 mm which represents an actual distance of 1 km. Find the scale to which the map is drawn.	_____ mm to _____ m
5	Write the part which is shaded (a) as a fraction (b) as a decimal (c) as a percentage.	(a) _____ (b) _____ (c) _____ %
6	Chloe gave a FIFTY and a TWENTY to pay for an item which cost 57p. How much change did she receive?	_____ p
7	A rectangle measures 10 cm by 6 cm. Find the length of another rectangle of the same area if its width is 4 cm.	_____ cm
8	18 × 56 = 1008 How many more than 1008 is 18 × 59?	_____
9	The circumference of the wheel is 248.4 cm. Find to the nearest m the distance travelled in making 100 turns.	_____ m
10	The mass of a parcel is 10 kg. 5% of its mass is for packing. Find in kg and g the mass of the contents.	_____ kg _____ g
11	10 articles cost £6·28. Find to the nearest penny the cost of one article.	_____ p
12	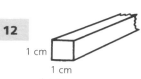 A metal rod is 1 cm square in section and 1 m long. (a) How many cm cubes can be cut from it? (b) Write the volume of the bar in cm³.	(a) _____ (b) _____ cm³

Section 3 Test 4

A

		ANSWER
1	2.5 m + 43 cm = ▢ cm	____ cm
2	£0·96 = 8 TWOS + ▢ TWENTIES	____ TWENTIES
3	Write in figures $\frac{1}{2}$ million.	____
4	$\frac{4}{5}$ = ▢ %	____ %
5	$\frac{5}{8} + \frac{1}{2} + 2$	____
6	1.35 litres = ▢ mℓ	____ mℓ
7	150 ÷ 8 = ▢ rem. ▢	____ rem. ____
8	4.38 × 6	____
9	£1·94 − 86p = £ ▢	____ £
10	5% of 300 g	____ g
11	$\frac{9}{10}$ of £4·00	____ £
12	1.050 km + ▢ m = 2 km	____ m

B

		ANSWER
1	Write as a decimal fourteen units plus seventeen thousandths.	____
2	How many pennies remain when £1·11 is divided by 9?	____
3	Find the product of 0.5 and 0.8.	____
4	What fraction in its lowest terms is (a) 40 min of 1 hour (b) 300 mℓ of $\frac{1}{2}$ litre?	(a) ____ (b) ____
5	(a) Write the date which is 7 months later than 1st September '96. (b) How many days are there in that month?	(a) ____ (b) ____
6	Find the cost of 1 kg 200 g at 50p per $\frac{1}{2}$ kg.	____ £
7	(a) 20% of £4·50 (b) 60% of £4·50	(a) ____ p (b) ____ £
8	How many times greater than 3.04 is 3040?	____
9	What speed in km/h is the same as $26\frac{1}{2}$ km in 15 min?	____ km/h
10	105° 70° 190° 175° 210° Which of the angles are reflex angles?	____° ____°
11	Approximate (a) 5050 to the nearest 100 (b) 29 632 to the nearest 1000.	(a) ____ (b) ____
12	The area of a rectangle is 60 m². Its length is 8 m. Find the width of the rectangle.	____ m

C

		ANSWER
1	The heights of 3 children are 140 cm, 160 cm and 135 cm. Find the average height in m and cm.	____ m ____ cm
2	A car travels 170 km on 20 litres of petrol. How many km per litre?	____ km
3	A letter was posted in Australia on 24th October and delivered in England on 4th November. For how many days was it in the post? Include the day of posting.	____
4	3 articles cost 27p. Find the cost of 7 articles.	____ p
5	Write each of these numbers so that the value of the figure 6 is 6 hundredths. (a) 306 (b) 463 (c) 2586	(a) ____ (b) ____ (c) ____
6	ABCD is a parallelogram and ∠ ABC is 55°. What is the size of ∠ ADC ∠ DAB ∠ BCD?	____° ____° ____°
7	A rectangular field 150 m wide required 800 m of fencing to enclose it. How long is the field?	____ m
8	A 5p coin weighs 3.25 g. By how many g is the mass of a £5 bag of FIVES greater than $\frac{1}{4}$ kg?	____ g
9	CAMERA £87 10% reduction for cash. Find (a) the reduction for cash (b) the cash price.	(a) £ ____ (b) £ ____
10	The area of a circle is 78.5 cm². Find the area of (a) the semicircle (b) the quadrant to the nearest cm².	(a) ____ cm² (b) ____ cm²
11	On a map the distance between two towns is 6.3 cm. If the map was drawn to the scale 1 cm to 5 km, find the actual distance between the towns.	____ km
12	(box 8 cm × 10 cm × 3 cm) (a) How many cm cubes are needed to fill the box? (b) If the box were 4 cm high, find its volume in cm³.	(a) ____ (b) ____ cm³

Turn back to page 30 and work for the second time Progress Test 2. Enter the result and the date on the chart.

Section 3 Test 5

A

		ANSWER
1	100 − 28 = 9 × ☐	
2	250 000 = ☐ million	
3	18p + 19p + 23p = £ ☐	£
4	1.2 m − 75 cm = ☐ cm	cm
5	450 g × 6 = ☐ kg	kg
6	(a) 10% of 840	(a)
	(b) 30% of 840	(b)
7	£3·70 ÷ 8 = ☐ p rem. ☐ p	p rem. p
8	3 min 50 s = ☐ s	s
9	$1 - \frac{3}{4} - \frac{1}{8}$	
10	3.0 × 0.8	
11	3.050 litres + ☐ mℓ = 4 ℓ	mℓ
12	(a) 0.55 = ☐ %	(a) %
	(b) $\frac{1}{25}$ = ☐ %	(b) %

B

		ANSWER
1	4 5 6 7 8 9 Which of these numbers are factors of 54?	
2	How many FIVES are there in £3·65?	
3	What is the average of 200 mℓ, 250 mℓ, 150 mℓ, 120 mℓ?	mℓ
4	Write each of these scores as a percentage (a) 5 out of 25	(a) %
	(b) 16 out of 50.	(b) %
5	How many m are there in 1.650 km?	m
6	Increase £50 by 25%.	£
7	How many h and min from 10.40 a.m. to 12.50 p.m.?	h min
8	$\frac{7}{10}$ of 5 m = 350 cm Find $\frac{3}{10}$ of 5 m.	cm
9	How many hundredths are there in three point nought four?	hundredths
10	200 g cost 38p. Find the cost of $\frac{1}{2}$ kg.	p
11	635 ÷ 7. Write the answer to the nearest whole one.	
12	Find the area of the triangle.	

C

		ANSWER
1	Find the difference between the sum of 6 and 7 and the product of 6 and 7.	
2	$\frac{3}{5}$ 0.03 $\frac{3}{10}$ $\frac{20}{50}$ 0.3 Which of these fractions are equivalent to 30%?	
3	Megan has 25p and Josh has 41p. How much must Josh give to Megan so that they have equal amounts?	p
4	A tin when $\frac{3}{4}$ full holds 720 mℓ. How many mℓ does it hold when it is (a) $\frac{1}{8}$ full	(a) mℓ
	(b) $\frac{3}{8}$ full?	(b) mℓ
5	A railway journey takes 1 h 38 min. If the train departs at 11.30 at what time does it arrive?	
6	A boy faces SW. In which direction is he facing if he turns (a) 90° clockwise	(a)
	(b) 45° anticlockwise?	(b)
7	(a) What is the name of this solid?	(a)
	(b) Find its surface area.	(b)
8	A customer paid 84p for 300 g of grapes. Find the price for $\frac{1}{2}$ kg.	£
9	Which of the following numbers do not change in value if the noughts are omitted? 0.158, 0350, 0.590, 1.506	
10	The perimeter of a rectangle is 54 cm. Its length is 18 cm. Find (a) its width	(a)
	(b) its area.	(b)
11	Find the smallest number which must be added to 403 to make it exactly divisible by 8.	
12	(a) What fraction of the circumference of the circle is the arc AB?	(a)
	(b) If the angle at the centre were 40°, what fraction of the circumference would be its arc?	(b)

Section 3 Test 6

A

1. Write as a decimal $10 + \frac{8}{100} + \frac{3}{1000}$.
2. 3.125 litres = ◻ ml _____ ml
3. $\frac{3}{10}$ of £1·80 _____ p
4. 1% of twenty thousand _____
5. 38 mm + 26 mm + 40 mm = ◻ cm _____ cm
6. $10^2 - 4^3$ _____
7. 0.02 × 50 _____
8. $\frac{1}{2}$ kg = 135 g + ◻ g _____ g
9. $\frac{3}{4}$ h − $\frac{2}{3}$ h = ◻ min _____ min
10. 47p × 8 = £ ◻ _____ £
11. 0.25 + ◻ = 0.365 _____
12. £35 ÷ 4 _____ £

B

1. How many hundreds are there in thirty thousand seven hundred?
2. 100 pencils cost £5·92. Find the cost of 25 pencils. _____ £
3. By how many degrees does the temperature rise from −10°C to 4°C? _____ °C
4. | 16 | 24 | 36 | 54 | 72 |

   Which of these numbers are multiples of both 6 and 8?
5. Find the average of $2\frac{1}{4}$, $1\frac{3}{4}$ and $3\frac{1}{2}$.
6. By how many g is 750 g less than 1 kg 150 g? _____ g
7. Write 35 eighths as
   (a) an improper fraction
   (b) a mixed number.   (a) _____ (b) _____
8. 20% of a sum of money is 49p. Find 100% of the money. _____ £
9. How many days are there in the seventh month of the year?
10. Write as a fraction in its lowest terms
    (a) 15 out of 40
    (b) 28 out of 32.   (a) _____ (b) _____
11. Approximate 17.850 litres to the nearest $\frac{1}{2}$ litre. _____ ℓ
12. Find in degrees
    ∠ x _____ °
    ∠ y. _____ °

C

1. The approximate population of a city is forty thousand more than $\frac{1}{2}$ million. What is the approximate population? _____
2. A girl collected 50p in her money box. She has 17 pennies, 9 TWOS and some FIVES. How many FIVES has she? _____ FIVES
3. 480 men, women and children went to a concert. From the diagram find how many
   (a) men   (a) _____
   (b) women   (b) _____
   (c) children   (c) _____
   were at the concert.
4. The perimeter of a regular octagon is 21.6 cm. Find in mm
   (a) the length of one side   (a) _____ mm
   (b) the length of a side of a regular hexagon of the same perimeter.   (b) _____ mm
5. The price of a ticket was increased from 50p to 60p. What is the increase
   (a) as a fraction   (a) _____
   (b) as a percentage of the original price?   (b) _____ %
6. How many packets each containing 300 g can be made from 2.5 kg? How many g are left? _____ rem. _____ g
7. 1 litre or 1000 cm³ of water has a mass of 1 kg.
   (a) How many mℓ have the same volume as 1 cm³?   (a) _____ mℓ
   (b) What is the mass of 1 cm³ of water?   (b) _____ g
8. A room is $1\frac{1}{4}$ times as long as it is wide. If the width is 6 m find the area of the room. _____ m²
9. In the triangle ABC find
   ∠ ABC _____ °
   ∠ BAC _____ °
   ∠ ACB. _____ °
10. Lawn sand is made from 3 parts coarse sand and 2 parts peat.
    (a) What percentage of the mixture is sand?   (a) _____ %
    (b) Find the mass of peat required to make 2 kg of the lawn sand.   (b) _____ g
11. Motor Museum ADMISSION £3·40 children − half price
    What is the total admission price for mother, father and two children? _____ £
12. The lawn measures 12 m by 7 m. The path around it is 1.5 m wide. Find the area of the whole garden. _____

Section 3 Test 7

A

1. Write in figures $\frac{1}{10}$ of 1 million.
2. 85 g × 100 = ▢ kg _____ kg
3. 1.06 − 0.79
4. 480 mm = ▢ m _____ m
5. $\frac{5}{6}$ of 30p _____ p
6. 905 ÷ 100
7. (a) 1% of £3·00 = ▢ p (a) _____ p
   (b) 9% of £3·00 = ▢ p (b) _____ p
8. £0·04 × 50 _____ £
9. $\frac{3}{4}$ h + 35 min = ▢ h ▢ min _____ h _____ min
10. 2.650 litres = ▢ mℓ _____ mℓ
11. 19p + 7p + 5p + 8p = £ ▢ _____ £
12. $1\frac{3}{4} \times 8$

B

1. Write as a decimal the sum of 9 tenths and 37 thousandths.
2. Find the two missing numbers in this series. 70, 7, ▢, 0.07, ▢
3. What is the average of 15 cm, $\frac{1}{4}$ m and 17 cm? _____ cm
4. Increase £7·50 by 10%. _____ £
5. By how many twelfths is $1\frac{5}{6}$ less than $2\frac{1}{4}$?
6. Write in 24-hour clock times
   (a) 18 min before noon (a)
   (b) $\frac{1}{2}$ h after 7.57 p.m. (b)
7. Find the difference between $\frac{1}{5}$ of 25 and $\frac{1}{3}$ of 45.
8. 5% of a sum of money is £0·40. Find the whole amount. _____ £
9. Approximate
   (a) 10.250 litres to the nearest litre (a) _____ ℓ
   (b) £439·87 to the nearest £. (b) _____ £
10. 17.1 cm A———B
    Find in mm the length of a line 15 mm shorter than AB. _____ mm
11. What is the cost of 800 g at 90p per kg? _____ p
12. The area of a rectangle is $66\frac{1}{2}$ cm². The width of the rectangle is 7 cm. Find its length. _____ cm

C

1. Write in words the number which is equal to $(3 \times 10^3) + (6 \times 10^2)$.
2. There are 60 cards in a packet. There are 12 packets and 17 odd cards. How many cards are there altogether?
3. A bill for £4·23 is paid with a £5 note. Name the four coins given as change. ___ p ___ p ___ p ___ p
4. ABC is an isosceles triangle. (angle C = 49°) Find the angle at B _____ °
   the angle at A. _____ °
5. 1 litre of water has the same volume as 1000 cm³. What are the volumes in cm³ of (a) 200 mℓ (a) _____ cm³
   (b) 1.7 ℓ? (b) _____ cm³
6. A plank of wood 7.5 m long is cut into two parts so that one part is four times as long as the other. Find the length of each part. ___ m ___ m
7. TRAIN TIMES
	Depart	Arrive
A	10.45	12.35
B	16.18	18.00

   Which train, A or B, is the quicker and by how many minutes? _____ by _____ min
8. The total mass of 12 parcels of equal mass is 5.4 kg. Find the mass of
   (a) 4 parcels in kg and g (a) _____ kg _____ g
   (b) 1 parcel in g. (b) _____ g
9. The rectangle X has two axes of symmetry. How many axes of symmetry has
   (a) the rhombus Y (a)
   (b) the equilateral triangle Z? (b)
10. The circumference of a wheel is 45 cm. How many times will it turn in going 450 m?
11. By how much is it cheaper to pay for an article priced at £25 with a discount of 10%, than a discount of 8p in the £? _____ p
12. Find the area of the shape. (19 cm, 8 cm, 15 cm) _____ cm²

38

Section 3   Test 8

A   ANSWER

1. Write in figures 1.1 million.
2. £0·81 = 2 TWENTIES + 3 TENS + ▢ p   p
3. 8.07 × 6
4. 1.8 km + 450 m = ▢ km ▢ m   km   m
5. 50% of 1.9 kg = ▢ g   g
6. 275 ÷ 7 = ▢ rem. ▢   rem.
7. $10^3 - 10^2$
8. (a) $\frac{3}{5}$ = ▢ %   (a)   %
   (b) $\frac{9}{10}$ = ▢ %   (b)   %
9. £10 − (2 × 46p) = £ ▢   £
10. $1\frac{3}{10} + \frac{2}{3} + \frac{7}{10}$
11. 1.5 litres − ▢ ml = 600 ml   ml
12. £0·17 × 8   £

B   ANSWER

1. Find in degrees the reflex angle to
   (a) 85° (b) 148°.   (a)   ° (b)   °
2. How many 50-g cans have a total mass of 2 kg?
3. Decrease £3·50 by 4p in the £.   £
4. Approximate
   (a) 9 l 870 ml to the nearest litre   (a)   l
   (b) 3.560 kg to the nearest $\frac{1}{2}$ kg.   (b)   kg
5. What was the date 6 months before 1st March 2000?
6. Divide nine hundred and seventy-two by nine.
7. (a) 5% of £6·00 = ▢ p   (a)   p
   (b) 15% of £6·00 = ▢ p   (b)   p
8. At 90 km/h what distance is travelled in 10 min?   km
9. Find the cost of 1.75 litres at 60p per l.   £
10. Divide a length of 48 cm into two pieces so that one is twice as long as the other.   cm   cm
11. What fraction in its lowest terms is
    (a) £27 of £36
    (b) £2·50 of £20?   (a)   (b)
12. Find the area of the triangle. (17 cm, 5 cm)

C   ANSWER

1. There were 3500 spectators at a football match. 7% were women. How many women were there?
2. A bottle holds 300 ml. Find in litres the contents of 12 bottles.   l
3. In the diagram there are 100 small squares. Find as a fraction in its lowest terms the part which is
   (a) shaded
   (b) unshaded
   (c) coloured.   (a)   (b)   (c)
4. 3.75 × 8 = 30 Write the answers to
   (a) 375 × 8   (a)
   (b) 3.75 × 80.   (b)
5. A sea cruise started on 25th August and ended on 7th September. For how many days did the cruise last?
6. A plan is drawn to the scale of 1 mm to 50 cm. What fraction represents this scale?
7. The average mass of 3 parcels is 6 kg. Two of the parcels have a mass of 4.6 kg and 6 kg. Find the mass of the third parcel.   kg
8. A sheet of plywood 26 cm by 8 cm is cut into strips 2 cm wide. Find the total length of the strips.   cm
9. 1 ml or 1 cm³ of water has a mass of 1 g. Find the mass of water in kg
   (a) in a can which holds $1\frac{3}{4}$ litres   (a)   kg
   (b) in a tank the volume of which is 6400 cm³.   (b)   kg
10. 7 articles cost £2·47. Find to the nearest penny the cost of one.   p
11. The diameter of this wheel is 50 cm. How far in m does the wheel travel for 1 turn? (C = πd, π = 3.14)   m
12. The price of $\frac{1}{2}$ kg of potatoes was increased from 30p to 36p. Find the increase as a percentage of the original price.   %

Turn back to page 30 and work for the third time Progress Test 2. Enter the result and the date on the chart.

Section 3 Test 9

A

		ANSWER
1	$5.305 = 5 + \dfrac{\square}{1000}$	$\dfrac{}{1000}$
2	1% of sixteen thousand	
3	4.375 kg = ☐ g	_____ g
4	£2 − (27p + 65p) = £ ☐	£
5	2 ℓ 450 mℓ − 0.5 ℓ = ☐ mℓ	mℓ
6	0.4 + ☐ = 0.476	
7	£1·76 × 5	£
8	0.75 km = ☐ m + 325 m	m
9	$\dfrac{x}{100} = 13.07$ Find x.	
10	38 min + 2 h + 47 min = ☐ h ☐ min	____ h ____ min
11	(a) 10% of £20·40	(a) £
	(b) $2\tfrac{1}{2}$% of £20·40	(b) £
12	$\tfrac{1}{6}$ of 45 cm = ☐ mm	mm

B

		ANSWER
1		5 6 3 7 9	
	Which of these numbers are factors of 75?	
2	Find the average of 3, 1.4, 2 and 2.6.	
3	Write (a) $\tfrac{47}{8}$ as a mixed number	(a)
	(b) $5\tfrac{5}{6}$ as an improper fraction.	(b)
4	What is the cost of 400 g at £1·50 per $\tfrac{1}{2}$ kg?	£
5	By how many is 450 × 1000 less than $\tfrac{1}{2}$ million?	
6	By how many degrees does the temperature fall from 8°C to −5°C?	°C
7	25% of my money is £35. Find the whole amount.	£
8	Approximate to the nearest whole number (a) 79.63	(a)
	(b) 12.475.	(b)
9	By how many mm is 95.4 cm less than 1 m?	mm
10	Find the least number of pennies which must be added to 92p to make the amount exactly divisible by 6.	
11	What percentage of 2 litres is (a) 400 mℓ	(a) %
	(b) 100 mℓ ?	(b) %

12 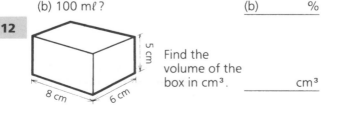 Find the volume of the box in cm³. _____ cm³

C

		ANSWER
1	By how many thousandths is 1.057 less than 2?	_____ thousandths
2	By how many degrees is the reflex angle AOB greater than the obtuse angle AOB?	°
3	A girl spends $\tfrac{1}{2}$ of her money on bus fares and $\tfrac{5}{12}$ on sweets. What fraction of her money is left?	
4	Find the distance travelled in 1 h 40 min at a speed of 90 km/h.	km
5	From the scale on the spring balance read as accurately as possible the mass shown by the pointer. Write the answer in g.	g
6	The population of a town was a quarter of a million, reduced later by 4%. By how many was the population reduced?	
7	How much change is there out of £15 after spending £9·50 and £3·20?	£
8	The total length of 5 laths is 16.25 m. Find in m and cm the length of 3 laths.	____ m ____ cm
9	A bus runs at intervals of 25 min. What are the times of the next two buses after 08.15?	____ ____
10	CASH PRICE £60 or 10% WEEKLY FOR 12 WEEKS — By how much is it cheaper to pay cash?	£
11	A path is 9 m long and 80 cm wide. Find its area in m².	m²
12	Ravi is given £13 as birthday gifts. He spends 9% and saves the remainder. (a) What percentage does he save?	(a) %
	(b) How much money does he spend?	(b) £

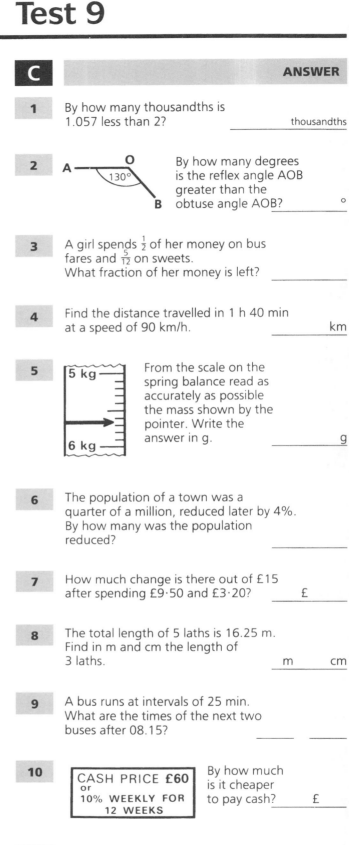

40

Section 3   Test 10

A

1. Write in figures six hundred and two thousand five hundred and eight.
2. £0·27 × 40   £
3. $\frac{2}{3}$ of 960
4. (a) 1% of £17   (a)   p
   (b) 7% of £17   (b) £
5. 10.06 = ◻ thousandths   thousandths
6. £2·30 = 3 FIFTIES + 2 TWENTIES + ◻ TENS   TENS
7. $3\frac{3}{4} \times 4$
8. 180° − (72° + 36°)   °
9. 300 g + 450 g + 350 g = ◻ kg   kg
10. 2 h 15 min − 50 min = ◻ h ◻ min   h   min
11. 27 cm × 7 = ◻ m ◻ cm   m   cm
12. $\frac{£16·56}{8}$ £

B

1. By how many is 0.3 million less than $\frac{1}{2}$ million?
2. What length in m is 6 times 7 m 30 cm?   m
3. How many times is $2\frac{1}{2}$ contained in 50?
4. Find the total of 17p, 53p, 24p and 7p. Write the answer in £s.   £
5. Write the 24-hour clock time which is 7 h before 4.35 a.m.
6. What fraction in its lowest terms is equal to (a) 8% (b) 35%?   (a)   (b)
7. By what quantity is 920 mℓ less than 1.2 litres?   mℓ
8. Increase £30·50 by 10%.   £
9. A ⊢— 4.7 km —⊣ B   C ⊢2.25 km⊣ D
   If the distance from A to D is 14 km, find the distance from B to C.   km
10. Find the cost of 0.3 m at £4·80 per m.   £
11. The perimeter of a rectangle is 65 cm. Its length is 24 cm. Find its width.   cm
12. 10 articles cost £2·09. Find the cost of one to the nearest penny.   p

C

1. How many articles costing 3p each can be bought for £2·40?
2. ABCD is a parallelogram. Find in degrees
   (a) ∠ BAD   (a)   °
   (b) ∠ ABC   (b)   °
   (angle 140° at D)
3. Share £5 between Josh and Amy so that Josh has 6p each time Amy has 4p. How much does each have?
   Josh £   Amy £
4. A motorist drives from Liverpool to Glasgow, a distance of 350 km, in 6 hours. Find to the nearest km his average speed in km/h.   km/h
5. | 0.33 | 0.3 | 0.03 | 0.333 |

   From the largest of these decimal fractions take the smallest.
6. 1 cm³ or 1 mℓ of water has a mass of 1 gram. A jar holds $3\frac{1}{2}$ ℓ of water. Find
   (a) the volume of water in cm³   (a)   cm³
   (b) the mass of the water in kg. (b)   kg
7. How much is saved by buying 15 kg at 16p per kg instead of the same mass at 9p per $\frac{1}{2}$ kg?   p
8. Find the age on the 1st Sept. 2010 in years and months of

   Dates of birth
   | Riaz | 1.3.'96 |
   | Sophie | 1.10.'94 |

   (a) Riaz   (a)   years   months
   (b) Sophie (b)   years   months
9. The area of a hall is 60 m². Its length is 8 m. Find (a) its width   (a)
   (b) its perimeter.   (b)
10. (a) What fraction of the circumference of the circle is the arc AB?   (a)
    (b) If the circumference measures 188.4 cm, find in mm the length of the arc.   (b)   mm
    (angle 120° at centre, A and B on circle)
11. Meat costs £2·60 per $\frac{1}{2}$ kg. Find the mass in kg and g of meat which costs £7·80.   kg   g
12. Box A measures 8 cm long, 9 cm wide, 4 cm high.
    Box B measures 10 cm long, $5\frac{1}{2}$ cm wide, 6 cm high.
    Find the difference in their volumes.

Section 3 Test 11

A

1. 10 × 10 × 10 × 10 × 10
2. 67p × 6 = £ ___   £ _____
3. (49 + 8) = 100 − ___
4. 90% of £300   £ _____
5. 5 litres ÷ 8 = ___ mℓ   mℓ _____
6. $10 - 7\frac{3}{10}$
7. 0.246 = 2 tenths + ___ thousandths   ___ thousandths
8. (a) $\frac{3}{25}$ = ___ %   (a) ___ %
   (b) 0.07 = ___ %   (b) ___ %
9. 1.25 kg − 600 g = ___ g   ___ g
10. 9 FIVES + 3 TWOS + 3 TWENTIES = £ ___   £ _____
11. $\frac{3}{4}$ of 3.6 cm = ___ mm   ___ mm
12. 2 h 49 min + 53 min = ___ h ___ min   ___ h ___ min

B

1. By how many is 90 200 less than one hundred thousand?
2. Find the total number of days in February, March and April in a leap-year.
3. How many degrees in the reflex angle AOB? (circle with angle 72° at O between A and B) ___ °
4. Find the cost of 90 cm of cloth at £3·70 per m.   £ _____
5. How many times is 400 g contained in 2.4 kg?
6. Of these numbers which is the smallest? 1.11, 1.01, 1.111, 1.1
7. What is the difference in mℓ between 2.8 ℓ and 3.7 ℓ?   ___ mℓ
8. Write as a fraction in its lowest terms
   (a) £4·50 of £18
   (b) 5 min of 1 h.   (a) ___ (b) ___
9. A strip of plastic 4 m 200 mm long is cut into 7 equal pieces. Find in mm the length of each piece.   ___ mm
10. Approximate
    (a) £29·50 to the nearest £   (a) £ ___
    (b) $\frac{99}{2}$p to the nearest penny.   (b) ___ p
11. 5)£x 93p   Find the sum of money which was divided by 5.   £ _____
12. The diameters of two circles are 9.4 cm and 15.8 cm. What is the radius of each circle in mm?   ___ mm ___ mm

C

1. Find the numbers less than 50 of which both 2 and 7 are factors. ___ ___ ___
2. What is the date of the third Wednesday in July if the 1st of July is on Sunday?
3.

June	July	Aug	Sept
40 mm	23 mm	42 mm	35 mm

The monthly rainfall is given in mm. Find the average rainfall for the 4 months.   ___ mm

4. 60 kg of mortar is mixed from 4 parts of sand and 1 part of cement. Find the mass used of
   (a) sand   (a) ___ kg
   (b) cement.   (b) ___ kg
5. Write this number in words. $(7 \times 10^3) + (1 \times 10^2) + (9 \times 10)$
6. How many degrees are there in a turn from W to SE (a) clockwise   (a) ___ °
   (b) anticlockwise?   (b) ___ °
7. Diameter of circle A is 9.4 cm. Diameter of circle B is 5.8 cm. How far apart are the two centres in mm?   ___ mm
8. (a) Find the area of a square of 5 cm side.   (a) ___
   (b) How many times greater is the area of a square with sides double that length?   (b) ___
9. STICKERS 7p each or 6 for 40p. How much money is saved by buying 24 stickers six at a time?   ___ p
10. 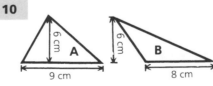 Which triangle A or B has the greater area and by how many cm²?   ___ by ___
11. A boy won a prize of £25 which he deposited in a bank at an interest rate of 6%. How much interest did he receive at the end of 1 year?   £ _____
12.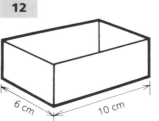
    (a) How many centimetre cubes can be fitted into the bottom of the box?   (a) ___
    (b) If the volume of the box is 240 cm³, find its height.   (b) ___ cm

Section 3 Test 12

A — ANSWER

1. $55 + 17 = \square \times 9$
2. £0·95 = 1 FIFTY + 1 TWENTY + ☐ FIVES — FIVES
3. $2.3\text{ m} - 90\text{ cm} = \square\text{ cm}$ — cm
4. 1 000 000 = ☐ thousands — thousands
5. = ☐ hundreds — hundreds
6. $380\text{ g} \times 9 = \square\text{ kg}$ — kg
7. $7 \times y = 7532$ Find y.
8. (a) 1% of £27 — (a) p
   (b) 8% of £27 — (b) £
9. $\frac{1}{2} + \frac{3}{8} + \frac{3}{4}$
   $2.050\ \ell + \square\ \text{m}\ell = 2\frac{1}{2}\ \ell$ — mℓ
10. $360° - (75° + 80° + 130°)$ — °
11. $\frac{5}{6}$ of £90 — £
12. 4.07×8

B — ANSWER

1. | | | | | |
   |---|---|---|---|---|
   | 24 | 36 | 54 | 60 | 72 |

   Which of the numbers are multiples of 4, 6 and 9?
2. How many km are there in seventeen hundred metres? — km
3. Decrease £44 by 10%. — £
4. How many h and min from 09.48 to 11.19? — h min
5. 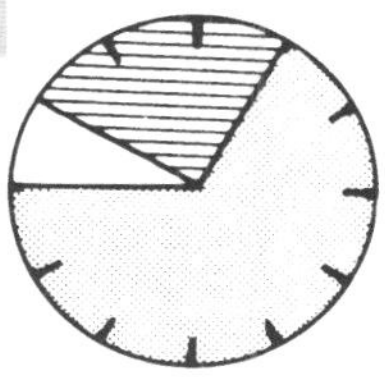

   What fraction in its lowest terms is
   (a) the part shaded — (a)
   (b) the part shaded ☰ — (b)
   (c) the part unshaded? — (c)
6. Find the cost of 2.25 ℓ at 28p per ℓ. — p
7. How many biscuits each costing 7p are bought for £2·73?
8. Write each score as a percentage.
   (a) 18 out of 20 — (a) %
   (b) 35 out of 35. — (b) %
9. 5 articles cost 80p. What fraction of 80p will 3 articles cost?
10. By how many is 300 050 greater than $\frac{1}{4}$ million?
11. Find $\frac{1}{5}$ of £1·68 to the nearest penny. — p
12. Find in cm³ the volume of a box 15 cm by 10 cm by 7 cm. — cm³

C — ANSWER

1. Find the missing numbers in this series.
   0.125, 0.25, 0.375, ,
2. What number when added to 48 three times gives a total of 120?
3. A date and raisin cake has a mass of 1.5 kg. If 40% of the mass is fruit, find the mass of the fruit in g. — g
4. A line 8 cm long is drawn to the scale 1 mm to 0.1 m. What length does the line represent? — m
5. | | | | | |
   |---|---|---|---|---|
   | $\frac{1}{3}$ | $\frac{3}{10}$ | $\frac{2}{5}$ | $\frac{1}{6}$ | $\frac{3}{8}$ |

   Which of these fractions is less than $\frac{1}{4}$?
6. Lawn sand is spread at the rate of 125 g per 1 m². How many kg are required to treat a lawn 50 m²? — kg
7. A motorist starts a journey of 36 km at 09.45 and arrives at 10.30. At this speed
   (a) how far does he travel in $\frac{1}{4}$ h? — (a) km
   (b) Find his average speed in km/h. — (b) km/h
8. The length of a rectangle is three times its breadth. If the perimeter is 192 cm find (a) the length — (a) cm
   (b) the breadth of the rectangle. — (b) cm
9. 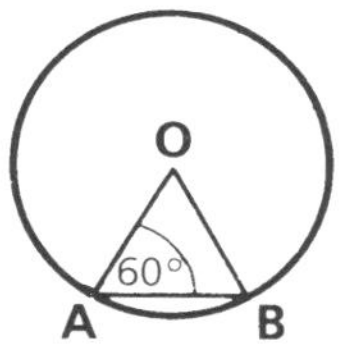

   O is the centre of the circle the radius of which is 7.4 cm. Find
   (a) the angle at the centre AOB — (a) °
   (b) the length of the straight line AB. — (b) cm
   (c) Name the triangle AOB according to its sides. — (c)
10. Of 150 children in a school 60 can swim 1 width of the pool, 45 can swim 1 length of the pool. What percentage of the children can swim (a) the width — (a) %
    (b) the length? — (b) %
11. 7 children shared a money prize equally. Each child received 42p and there was 6p left. Find the total value of the prize. — £
12. 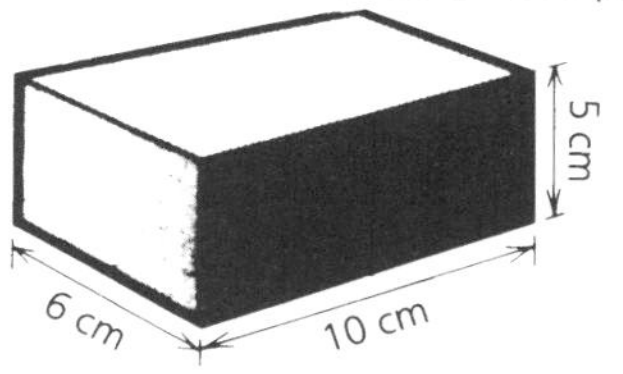

    The drawing shows a block of metal.
    (a) Find its volume.
    (b) If the mass of the metal is 7 times that of water, what is the mass of the block in kg? — kg

Turn back to page 30 and work for the fourth time Progress Test 2. Enter the result and the date on the chart.

CHECK-UP TEST Number

A

(a)
- 5 + 6
- 8 + 8
- 0 + 7
- 7 + 8
- 4 + 7
- 18 + 9
- 15 + 8
- 3 + 29
- 7 + 36
- 14 + 19
- 12 − 5
- 9 − 0
- 11 − 3
- 14 − 5
- 15 − 9
- 24 − 6
- 26 − 9
- 32 − 8
- 58 − 9
- 47 − 20

(b)
- 10 × 10
- 4 × 7
- 9 × 3
- 8 × 6
- 1 × 8
- 5 × 9
- 7 × 7
- 0 × 0
- 4 × 8
- 9 × 7
- 24 ÷ 3
- 40 ÷ 8
- 0 ÷ 6
- 54 ÷ 9
- 7 ÷ 7
- 42 ÷ 7
- 81 ÷ 9
- 36 ÷ 4
- 63 ÷ 9
- 56 ÷ 8

B

- (6 × 6) + 5
- (9 × 1) + 7
- (5 × 8) + 4
- (8 × 0) + 6
- (10 × 5) + 8
- (8 × 8) + 6
- (3 × 3) + 2
- (9 × 8) + 7
- (4 × 9) + 5
- (7 × 6) + 3
- 29 ÷ 3 ___ rem.
- 67 ÷ 8 ___ rem.
- 21 ÷ 4 ___ rem.
- 6 ÷ 7 ___ rem.
- 39 ÷ 5 ___ rem.
- 70 ÷ 9 ___ rem.
- 51 ÷ 6 ___ rem.
- 13 ÷ 7 ___ rem.
- 52 ÷ 5 ___ rem.
- 4 ÷ 9 ___ rem.

C

- 27 × 8
- 49 × 6
- 107 × 7
- 93 × 10
- 180 × 10
- 95 × 20
- 86 × 40
- 100 × 80
- 98 × 100
- 204 × 100
- 102 ÷ 3
- 336 ÷ 4
- 648 ÷ 6
- 590 ÷ 10
- 800 ÷ 10
- 540 ÷ 20
- 420 ÷ 60
- 1050 ÷ 50
- 4000 ÷ 100
- 2900 ÷ 100

D Write these numbers.

- Fifty thousand and seven
- Sixty-two thousand four hundred and two
- One hundred and forty thousand and eleven
- Two hundred and six thousand and nine
- 30 000 + 400 + 6
- 100 000 + 7000 + 50 + 8
- (4 × 1000) + (6 × 100) + (3 × 10) + 8
- (9 × 1000) + (7 × 10) + 5
- (3 × 1000) + (4 × 10)
- 1 million
- 1½ million
- ¼ million
- 2.7 million

E Write as decimals.

- 47 tenths
- 201 tenths
- 4 hundredths
- 309 hundredths
- 580 hundredths
- 603 thousandths
- 75 thousandths
- 3009 thousandths
- $9 + \frac{3}{10} + \frac{8}{100}$
- $10 + \frac{7}{100} + \frac{2}{1000}$
- 5 tenths + 2 hundredths
- 17 hundredths and 6 thousandths

F How many **tenths** equal

- 6.8
- 14.9
- 30.4?

How many **hundredths** equal
- 0.93
- 7.05
- 3.2?

How many **thousandths** equal
- 0.003
- 0.078
- 1.52
- 2.8
- 4.09?

G

- 5.03 + 0.7
- 2.5 + 1.54
- 0.06 + 1.04
- 3.7 + 0.35
- 0.28 + 1.625
- 2 − 1.4
- 1.4 − 0.9
- 10 − 8.75
- 4.8 − 3.76
- 0.7 − 0.58

H

- 6.45 × 10
- 0.873 × 10
- 2.03 × 100
- 0.092 × 100
- 1.64 × 1000
- 0.053 × 1000
- 1.8 × 5
- 4 × 1.63
- 0.09 × 8
- 7 × 2.08
- 1.063 × 6

I

- 79 ÷ 10
- 40.2 ÷ 10
- 34 ÷ 100
- 10.7 ÷ 100
- 608 ÷ 1000
- 1035 ÷ 1000
- 5.6 ÷ 8
- 10.25 ÷ 5
- 0.636 ÷ 6
- 4.77 ÷ 9
- 8.032 ÷ 8

J Find the value of x.

- $x + 7 = 24$
- $5 + x = 32$
- $x + 1.5 = 5$
- $31 - x = 16$
- $x - 6.3 = 10$
- $10 \times x = 25$
- $x \times 4 = 18$
- $7 = \frac{x}{5}$
- $\frac{x}{10} = 0.6$
- $9 + x = 7 \times 7$

44

CHECK-UP TEST — Money and Measures

A
- 70p = £_____
- 2p = £_____
- £0·63 = _____p
- £0·19 = _____p
- £0·04 = _____p
- £1·37 = _____ TENS _____p
- £3·09 = _____ TWENTIES _____p
- £10·80 = _____ TENS _____p

- 7 TENS + 6 TWOS _____p
- 3 FIFTIES + 9 TENS £_____
- 3 TENS + 5 FIVES + 9p _____p
- £0·85 = 5 TENS + ___ FIVES
- £1·20 = 12 FIVES + ___ TWENTIES
- £2·30 = 3 FIFTIES + ___ TWENTIES

B
- 9p + 3p + 17p = _____p
- 15p + 8p + 6p = _____p
- 14p + 7p + 12p = _____p
- 5p + 11p + 15p + 4p = _____p
- 6p + 19p + 21p + 18p = _____p
- 37p + 85p = £_____
- £1·03 + 49p = £_____
- £2·57 + £0·60 = £_____

- 43p − 19p = _____p
- 95p − 18p = _____p
- £1·10 − 84p = _____p
- £1·70 − 93p = _____p
- £2·30 − £0·80 = £_____
- £2·06 − £1·40 = _____p

C Find the cost of
- 10 @ 15p each £_____
- 100 @ 3p each £_____
- 9 @ 13p each £_____
- 8 @ 27p each £_____
- 5 @ 45p each £_____
- 19 @ 4p each _____p
- 27 @ 7p each. £_____

Find the cost of 1 when
- 10 cost £2·70 _____p
- 100 cost £15 _____p
- 6 cost 84p _____p
- 4 cost £0·72 _____p
- 7 cost £2·24 _____p
- 9 cost £3·06. _____p

D Find the change from
- 20p after spending (a) 3p _____p (b) 8p _____p
- 20p after spending (a) 12p _____p (b) 14p _____p
- 50p after spending (a) 37p _____p (b) 19p _____p
- (c) 26p _____p (d) 5p _____p
- £1 after spending (a) 81p _____p (b) 66p _____p
- (c) 45p _____p (d) 7p _____p
- £5 note after spending (a) 73p £_____ (b) £4·09 _____p
- (c) £2·54 £_____ (d) £1·98 £_____

E Make up the given amounts using the least number of coins. The first one is done for you.

AMOUNT	50p	20p	10p	5p	2p	1p
23p			1		1	1
39p						
67p						
78p						
86p						
94p						

F
- 84 cm = _____ m
- 309 cm = _____ m
- 1075 mm = _____ m
- 2305 mm = _____ m
- 750 mm = _____ m
- 100 m = _____ km
- 925 m = _____ km
- 1605 m = _____ km
- 860 g = _____ kg
- 1400 g = _____ kg
- 700 mℓ = _____ ℓ
- 3310 mℓ = _____ ℓ

G
- 20.4 cm = _____ mm
- 1.5 m = _____ mm
- 2.65 m = _____ mm
- 0.85 m = _____ cm
- 8.37 km = _____ m
- 0.6 km = _____ m
- 10.075 km = _____ m
- 1.325 kg = _____ g
- 0.05 kg = _____ g
- 3.72 kg = _____ g
- 1.3 ℓ = _____ mℓ
- 4.25 ℓ = _____ mℓ

H Find the cost of
- 500 g @ 76p per kg _____p
- 100 g @ 50p per kg _____p
- 250 g @ 36p per kg _____p
- 200 g @ £1·20 per kg _____p
- 1.5 kg @ 64p per kg _____p
- 100 g @ 45p per ½ kg _____p
- 300 g @ £1·10 per ½ kg _____p
- 25 cm @ 92p per m _____p
- 10 cm @ £3·50 per m _____p
- 60 cm @ £2·20 per m £_____
- 1.3 ℓ @ 60p per ℓ _____p
- 800 mℓ @ 50p per ℓ. _____p

I How many
- min in ¾ h _____
- min in 1¼ h _____
- s in 5 min _____
- weeks in 1 year _____
- days in 1 year _____
- days in April _____
- days in July _____
- days in October? _____

J Change to 24-hour clock times.
- 7.35 a.m. _____
- 12.05 p.m. _____
- 3.27 p.m. _____
- 10.55 p.m. _____

Change to 12-hour clock times. Use a.m. or p.m.
- 09.20 _____
- 14.56 _____
- 00.35 _____
- 21.16 _____

K Find the period of time between
- 8.35 a.m. and 10.16 a.m. ___ h ___ min
- 5.25 a.m. and noon ___ h ___ min
- 4.30 p.m. and 7.20 p.m. ___ h ___ min
- 11.35 and 14.15 ___ h ___ min
- 03.40 and 06.10. ___ h ___ min

How many days inclusive
- from 28th Jan. to 9th Feb. _____
- from 17th May to 5th June _____
- from 26th Nov. to 3rd Jan.? _____

CHECK-UP TEST — Fractions and Percentages

A
Fill in the missing numerator or denominator.

$\frac{3}{4} = \frac{\ }{16}$ $\frac{2}{3} = \frac{8}{\ }$ $\frac{7}{8} = \frac{\ }{24}$ $\frac{5}{6} = \frac{\ }{18}$ $\frac{4}{5} = \frac{40}{\ }$ $\frac{3}{10} = \frac{30}{\ }$

Reduce each fraction to its lowest terms.

$\frac{9}{12} = \underline{\ \ }$ $\frac{12}{18} = \underline{\ \ }$ $\frac{20}{25} = \underline{\ \ }$ $\frac{24}{30} = \underline{\ \ }$ $\frac{70}{100} = \underline{\ \ }$ $\frac{45}{100} = \underline{\ \ }$

Change each improper fraction to a mixed number.

$\frac{19}{4} = \underline{\ \ }$ $\frac{31}{5} = \underline{\ \ }$ $\frac{43}{8} = \underline{\ \ }$ $\frac{29}{6} = \underline{\ \ }$ $\frac{77}{10} = \underline{\ \ }$ $\frac{40}{3} = \underline{\ \ }$

Change each mixed number to an improper fraction.

$7\frac{3}{4} = \underline{\ \ }$ $8\frac{2}{3} = \underline{\ \ }$ $5\frac{4}{5} = \underline{\ \ }$ $9\frac{7}{10} = \underline{\ \ }$ $4\frac{7}{8} = \underline{\ \ }$ $10\frac{5}{6} = \underline{\ \ }$

B
Write as a fraction in its lowest terms.

50 of 75 _____
30p of £1·00 _____
25 cm of 1 m _____
12 kg of 30 kg _____
70 of 100 _____
800 g of 1 kg _____
400 mℓ of 2 ℓ _____
45 of 100 _____

C
Find

$\frac{3}{5}$ of 70 _____
$\frac{5}{8}$ of 64 _____
$\frac{7}{10}$ of £1·20 _____ p
$\frac{5}{6}$ of 42 ℓ _____ ℓ
$\frac{4}{7}$ of 350 g _____ g
$\frac{13}{100}$ of £1·00 _____ p
$\frac{2}{3}$ of 1200 _____
$\frac{35}{100}$ of 1 kg. _____ g

D
Find the whole when

$\frac{1}{6}$ is 35 _____
$\frac{3}{4}$ is 27p _____ p
$\frac{4}{5}$ is 36 cm _____ cm
$\frac{7}{10}$ is £1·40 _____ £
$\frac{2}{3}$ is 800 g _____ g
$\frac{5}{9}$ is 5000 _____
$\frac{3}{8}$ is 24 ℓ _____ ℓ
$\frac{9}{20}$ is £1·80 _____ £

E
Write as percentages.

(a) 33 out of 100 ___ %   (b) 87 out of 100 ___ %   (c) 9 out of 100 ___ %   (d) 45 out of 100 ___ %

(a) 0.65 ___ %   (b) 0.38 ___ %   (c) 0.75 ___ %   (d) 0.3 ___ %

(a) $\frac{29}{100}$ ___ %   (b) $\frac{56}{100}$ ___ %   (c) $\frac{1}{100}$ ___ %   (d) $\frac{13}{100}$ ___ %

Change each fraction first to hundredths, then write it as a percentage.

(a) $\frac{19}{50} = \frac{\ }{100} = \underline{\ \ }$ %   (b) $\frac{3}{25} = \frac{\ }{100} = \underline{\ \ }$ %   (c) $\frac{13}{20} = \frac{\ }{100} = \underline{\ \ }$ %

(a) $\frac{3}{4} = \frac{\ }{100} = \underline{\ \ }$ %   (b) $\frac{4}{5} = \frac{\ }{100} = \underline{\ \ }$ %   (c) $\frac{7}{10} = \frac{\ }{100} = \underline{\ \ }$ %

Fill in the blank spaces in each of the columns marked a – m. The first is done for you.

	(a)	(b)	(c)	(d)	(e)	(f)	(g)	(h)	(i)	(j)	(k)	(l)	(m)
VULGAR FRACTION (LOWEST TERMS)	$\frac{1}{2}$			$\frac{1}{5}$				$\frac{1}{10}$			$\frac{1}{20}$		
DECIMAL FRACTION	0.5	0.25			0.4		0.8		0.3		0.9		0.01
PERCENTAGE	50%	%	75%	%	%	60%	%	%	%	70%	%	%	%

F
Find the value of

25% of 120 _____
50% of 35 _____
75% of 400 _____
10% of 1000 _____
30% of 90 _____
70% of 200 _____
90% of 160 _____
20% of 95p _____ p
40% of £20 _____ £
60% of £15. _____ £

G
Find the value of

50% of 14p _____ p
20% of £6·50 _____ £
100% of 93p _____ p
10% of 2.5 kg _____ g
5% of 4 ℓ _____ mℓ
30% of 2 m _____ cm
1% of £1·00 _____ p
7% of £1·00 _____ p
3% of £3·00 _____ p
12% of £9·00. _____ £

H
Find as a percentage

6 of 24 _____ %
$7\frac{1}{2}$ of 15 _____ %
40p of 50p _____ %
93p of 93p _____ %
200 g of $\frac{1}{2}$ kg _____ %
700 mℓ of 1 ℓ _____ %
25p of £2·50 _____ %
£1·50 of £2·00 _____ %
7p of £1·00 _____ %
30 cm of 1.5 m. _____ %

CHECK-UP TEST — Approximations, Angles and Shapes

A Approximate to the nearest

whole number	49.55	_____
whole number	$20\frac{2}{5}$	_____
hundred	6057	_____
hundred	19 503	_____
thousand	59 770	_____
thousand	109 495	_____
£1·00	£27·50	£ _____

Find to the nearest penny (a) $\frac{1}{10}$ of 97p _____ p

B Approximate to the nearest

metre	8 m 59 cm	_____ m
metre	19 m 700 mm	_____ m
kilogram	16 kg 50 g	_____ kg
kilogram	7.550 kg	_____ kg
$\frac{1}{2}$ kg	9 kg 800 g	_____ kg
$\frac{1}{2}$ kg	6.550 kg	_____ kg
litre	39.870 ℓ.	_____ ℓ

(b) $\frac{1}{3}$ of £2·50 _____ p   (c) $\frac{£3·35}{4}$ _____ p

C How many degrees in each of the angles x and y?

angle x _____ °

angle x _____ °
angle y _____ °

angle x _____ °
angle y _____ °

D Find the missing angle in each of the triangles. Then name each triangle according to (a) the angles (b) the sides.

ANGLES IN TRIANGLE			(a) NAME OF TRIANGLE (ANGLES)	(b) NAME OF TRIANGLE (SIDES)
32°		90°		
46°	52°			
60°		60°		
	125°	38°		
57°	57°			

E Find the angle x in

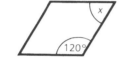

(a) the rhombus _____ °   (b) the trapezium _____ °
(c) the parallelogram _____ °   (d) the irregular quadrilateral _____ °

F Give the unit of measurement in the answer for each example.

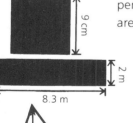

perimeter of square _____
area of square _____
perimeter of rectangle _____
area of rectangle _____
area of triangle _____
diameter of circle _____
circumference of circle _____
$\{ C = \pi d \text{ or } 2\pi r \}$
$\{ \pi = 3.14 \}$

How many cm cubes
(a) fit into the bottom of the box _____
(b) fill the box? _____

Write the missing measurement in each of the squares or rectangles.

Area	50 m²		25 cm²	16 m²
Length		13.5 cm		
Breadth	10 m	9 cm	2.5 cm	50 cm

Write the missing measurement in each of the triangles.

Base	16 cm	45 m	10 cm	
Height	8 cm	12 m		7 cm
Area			90 cm²	42 cm²

Write the missing radius or diameter.

radius	15.3 cm			27.6 cm
diameter		36 mm	9.8 cm	

Find the circumference of a circle when
d = 7 cm _____   r = 4.5 cm _____

Find the volume of each of these boxes.
length 13 cm, breadth 8 cm, height 2 cm _____
length 7 cm, breadth 4 cm, height 2.5 cm _____
cube of 6 cm side _____

Schofield & Sims

Schofield & Sims has, for over a hundred years, published a wide variety of educational materials for use in school. Specialising in products for early years, Key Stage 1 and Key Stage 2, our texts are written by experienced teachers and concentrate on the key areas of maths, English and science.

Mental Arithmetic 4

This book is one of a series providing carefully graded questions that develop pupils' essential mathematics skills and prepare them for all aspects of the Key Stage 2 maths tests, including mental maths. *Schofield & Sims Mental Arithmetic* is designed primarily for pupils in Key Stage 2 who are working at levels 3 to 6. However, the Introductory Book may also be suitable for use at Key Stage 1, and books 5 and 6 provide a bridge to Key Stage 3.

Every *Schofield & Sims Mental Arithmetic* book is divided into three sections, each comprising 12 one-page tests presented in a standard format that pupils will quickly become familiar with. Every test in **Mental Arithmetic 4** contains:

- **Part A:** 12 questions where use of language is kept to a minimum – based on the signs = ,+, –, x and ÷
- **Part B:** 12 questions using number vocabulary – particularly the language associated with the four signs
- **Part C:** 12 questions presented in word form as one- or two-stage problems.

A useful **Language of Maths** glossary on the inside front cover helps to develop pupils' number vocabulary. Two 10-minute **Progress Tests** are provided, with accompanying **Results Charts**. Final **Check-up Tests** (on number, money, measures, fractions, percentages, approximation, angles and shapes) help pupil and teacher to identify any gaps in understanding.

Also available: the remaining pupil books as listed below. Separate books of answers – for quick and easy marking – may also be obtained from Schofield & Sims.

Mental Arithmetic Introductory Book	Mental Arithmetic 3 ISBN 978 07217 0801 0
(for Key Stages 1 and 2) ISBN 978 07217 0798 3	Mental Arithmetic 5 (for Key Stages 2 and 3) ISBN 978 07217 0803 4
Mental Arithmetic 1 ISBN 978 07217 0799 0	Mental Arithmetic 6 (for Key Stages 2 and 3) ISBN 978 07217 0804 1
Mental Arithmetic 2 ISBN 978 07217 0800 3	

I can do maths

Many schools are now using *Schofield & Sims Mental Arithmetic* with the **I can do** teaching method as a whole-school differentiated maths programme. For full details, order the **I can do maths Teacher's Guide** (ISBN 978 07217 1115 7) from Schofield & Sims, using the contact details below.

Other Key Stage 2 maths materials available from Schofield & Sims include:

Understanding Maths – eight learning workbooks for Key Stage 2
Key Stage 2 Maths Revision Guide – comprehensive coverage of every topic likely to be covered in the Key Stage 2 tests
Key Stage 2 Maths Practice Papers – with each page cross-referenced to the Revision Guide.

This edition copyright © Schofield and Sims Ltd., 2007
Reprinted 2007 (twice), 2008
First edition compiled by J. W. Adams and R. P. Beaumont, and edited by T. R. Goddard; copyright © Schofield & Sims Ltd., 1976
British Library Cataloguing in Publication Data: A catalogue record for this book is available from the British Library.

All rights reserved. Except where otherwise indicated, no part of this publication may be reproduced or transmitted in any form or by any means, electronic or mechanical, including photocopying, recording or duplication in any information storage and retrieval system, without permission in writing from the publisher.
The Progress Tests and Results Charts may be photocopied after purchase for use within your school or institution only. This publication is not otherwise included under the licences issued by the Copyright Licensing Agency Ltd., Saffron House, 6-10 Kirby Street, London EC1N 8TS.

Front cover design by Curve Creative, Bradford; redesign of selected pages at reprint by Ledgard Jepson, Sheffield, and Wyndeham Gait Ltd., Grimsby
Printed in the UK by Wyndeham Gait Ltd., Grimsby, Lincolnshire

Schofield & Sims

ISBN 978-07217-0802-7

Dogley Mill, Fenay Bridge, Huddersfield HD8 0NQ
Phone: 01484 607080 Facsimile: 01484 606815
E-mail: sales@schofieldandsims.co.uk
www.schofieldandsims.co.uk
For information on **I can do maths**, visit
www.schofieldandsims.co.uk/icando/

ISBN 978 07217 0802 7
£2.45
(Retail price)
Key Stage 2
Age range 7–11 years